Shifting Cultivation Systems

Shri Kant Tripathi · Francis Q. Brearley
Editors

Shifting Cultivation Systems

From Indigenous Practices to Soil Fertility in North-East India

Springer

Editors
Shri Kant Tripathi
Department of Forestry
Mizoram University
Aizwal, India

Francis Q. Brearley
Department of Natural Sciences
Manchester Metropolitan University
Manchester, UK

ISBN 978-3-031-70387-4 ISBN 978-3-031-70388-1 (eBook)
https://doi.org/10.1007/978-3-031-70388-1

This Springer imprint is published by the registered company Springer Nature Switzerland AG
The registered company address is: Gewerbestrasse 11, 6330 Cham, Switzerland

Contents

Introduction to Shifting Cultivation Systems in North-East India: From Indigenous Practices to Soil Fertility

Francis Q. Brearley and Shri Kant Tripathi

Shifting cultivation (also known as swidden or slash-and-burn agriculture) has been conducted by numerous rural societies across the moist tropics [1–3] supporting the livelihoods and food security options of millions of people globally [4]. This practice involves the selection of a piece of suitable forested land for cultivation, followed by slashing of the vegetation and burning the biomass in situ after it is dried. After biomass burning, the area is cultivated for one or a few years depending on the pressure from weeds and retention of soil fertility (Fig. 1). When the soil fertility and crop productivity decline, the land is abandoned allowing the fallow period of secondary forest regeneration to suppress weed populations and restore soil fertility through natural recovery of vegetation to compensate nutrients lost during previous harvests. As part of a landscape mosaic with sufficiently long fallow periods, the practice remains sustainable [5–7] and can contribute to biodiversity conservation and ecosystem service provision through retaining a complex landscape mosaic [7, 8].

North-east India consists of eight small states (often referred to as the 'Seven Sisters', along with their 'brother' Sikkim), with studies in this book particularly focussed on the states of Mizoram and Nagaland. In this sub-tropical region of India, shifting agriculture, practiced for centuries, serves as the foundation for the livelihoods of rural communities. It is colloquially known as 'jhum' agriculture in this region. Agro-biodiversity in the shifting cultivation fields can be remarkably high, particularly when considering indigenous landraces [9]. The system has been well studied in this region since the pioneering work by Ramakrishnan and students [10] who promoted the sustainability of this practice.

F. Q. Brearley (✉)
Department of Natural Sciences, Manchester Metropolitan University, Chester Street, Manchester M1 5GD, UK
e-mail: f.q.brearley@mmu.ac.uk

S. K. Tripathi
Department of Forestry, Mizoram University, Tanhril, Aizawl, Mizoram 796009, India

S. K. Tripathi and F. Q. Brearley (eds.), *Shifting Cultivation Systems*,
https://doi.org/10.1007/978-3-031-70388-1_1

Fig. 1 Typical shifting cultivation fields in Nagaland, North-east India (photo by Etsoshan Ovung)

However, shifting cultivation has long been subjected to criticism and prejudice, with this criticism crystalised by the FAO in 1957 [11], whilst government bodies in North-east India have introduced many policies to promote land-uses other than shifting cultivation, albeit with varying success [12, 13]. Nevertheless, shifting cultivation is based on sound agro-ecological practices and, previously, when the length of the fallow period was sufficiently prolonged to match the nutrients lost during crop harvest through forest regeneration (i.e. 20–30 years), the ecosystem quality was maintained, and the system was sustainable. Yet, in recent years, almost all publications on this topic highlight a decrease in the duration of fallow periods (i.e. 5 years or less), combined with an increased number of cycles. This is attributed to a growing population and an increased demand for food combined with a shrinking area available for cultivation due to forest loss for timber and/or plantations.

Whether or not fallow periods are declining further (e.g. [14]), it is clear that shorter fallow periods do not allow the soil to restore adequate fertility to support new crops [15, 16] and this has led to widespread concern about declines in soil fertility, crop yields, food security for the rural poor, and ecological sustainability of the region. Therefore, there is a need to formulate techniques with implications for nutrient management and the sustainability of shifting cultivation in the region. It has been noted that there are two main, non-exclusive methods by which this could be done [17]. Firstly, through more 'effective' fallows—whereby the efficiency of fallow functions is improved, such that similar production benefits can be realised over a shorter time frame. Secondly, via more 'productive' fallows—in which fallow lengths stay the same and value is added to the fallow by introducing perennial economically-valuable species. The importance of improving the efficiency of crop growth for increased food production is a third option that can be explored further in the light of nutrient management techniques.

This book aims to better understand nutrient management under shifting cultivation systems through exploring, contrasting, and complementing traditional approaches with 'cutting edge' scientific analytical techniques. It is an outcome of two workshops held at Mizoram University in North-east India funded by the UK Global Challenges Research Fund that focused on discussing various techniques and sharing experiences among researchers working on shifting cultivation, soil science and land-use change. In a nutshell, this book focuses on the steeply sloping shifting cultivation fields of North-east India where problems of soil erosion and nutrient losses are prevalent, and therefore recommendation of practices in this book were based considering these aspects.

As shifting cultivation practices are closely tied to the cultural life of Indigenous groups in North-east India, this is one of the reasons it has been retained in these societies. In the first chapter of the book, Changkija [18] outlines how the festivals and ceremonies celebrated by the Ao-Naga of Nagaland are linked to the various stages of the shifting cultivation calendar. Tripathi et al. [19] then follow by describing the practice of shifting cultivation as practiced by the Indigenous population of Mizoram and note how this has been altered in more recent years. This is followed by Changkija et al. [20] who provide a broader geographical view of the diversity of agricultural practices in the region with a focus on practices other than shifting cultivation but that make efficient use of natural resources. Giri et al. [21] examine the alder-based shifting cultivation practices in Konoma, Nagaland, where fallows are enriched by the nitrogen-fixing bacteria associated with alder roots. A series of chapters then look at aspects of soil science, soil nutrition and soil microbiology. Mishra et al. [22] show how soil characteristics influence changes in nutrient concentrations after shifting cultivation, suggesting that certain soil types may be more susceptible to degradation after such cultivation. Hombegowda et al. [23] look at soil organic carbon recovery in shifting cultivation fallows showing how it appears more rapid in North-east India compared to the Indian Eastern Ghats. Ovung et al. [24] examine changes in soil carbon and nitrogen stocks under shifting cultivation and show that they did not differ when compared with an undisturbed forest, and Mishra and Sarkar [25] explore how local farmers classify and choose soils for cultivation and how this classification agrees well with methods of soil classification by soil scientists. Ray et al. [26] and Luangmuana et al. [27] examine shifting cultivation practices from a more agronomic perspectives and provide their suggestions for ways forward for improving aspects of the practice. In terms of soil microbiology, Brearley et al. [28] examine how soil bacterial communities change during the fallow period of shifting cultivation showing that it takes at least ten years for bacterial communities to recover, which is similar to minimum suggested fallow periods in the region. Momin and Tripathi

[29] document some actinomycetes from shifting cultivation fields with the potential to act as bioinoculants, and Mehetre et al. [30] review the topic of soil microbial diversity in North-east Indian shifting cultivation systems presenting an extensive range of studies on various microbial taxa. Debnath et al. [31] and Tripathi et al. [32] complete the book with a pair of reviews examining the state of current research, and ways forward for shifting cultivation research and policy in North-east India.

Globally, tropical landscapes are changing, with shifting cultivation persisting in areas with low population densities and fewer opportunities for alternative agricultural and/or economic options [1]. There is now a slow transition away from shifting cultivation globally towards plantations and permanent cropping systems [33] but, given that shifting cultivation is an ‘organic-like’ agro-ecological farming technique [34], it has the potential to support the UN’s Sustainable Development Goals, ‘No Poverty’, ‘Zero Hunger’ and particularly ‘Life on Land’. Indeed, when making comparisons with other land-uses, it is actually most appropriate to compare the ecosystem services provided by the shifting cultivation fallows as this is the dominant land-use type [35]. This cultivation practice has the opportunity to support various aspects of biodiversity, including traditional crop varieties, and local cultural traditions and knowledge. However, that is not to say that improvements to the systems cannot be made in line with local conditions and farmer preferences. Various techniques have been suggested to provide future options for improving aspects of shifting cultivation whilst retaining its organic and sustainable nature. These include incorporating inter-row cropping between contour hedgerows of nitrogen-fixing shrubs, slope terracing, judicious mulching, and agroforestry with anti-erosional plants. However, some of these may catalyse a move towards a more permanent and settled farming system so need to be considered if they are appropriate in all cases. Additionally, litter amendments to soil from adjoining forest, microbial inoculants and enhanced fallow recovery rates using fast-growing species have also been suggested among the promising options to manage soil fertility and crop productivity in the region. Further consideration should also be made to incorporate aspects of integrated farming systems as there are many synergies to be gained between these two cultivation systems.

In conclusion, we, along with other authors [5–8, 10, 12, 36, 37], note that shifting cultivation has a role to play in food security, biodiversity conservation, preservation of ecosystem services, and supporting sustainable and culturally relevant livelihoods in many circumstances. Nevertheless, we also support providing options to shifting cultivation farmers for site-specific indigenous tree-based subsistence farming systems to develop resource use approaches through better tree-crop interactions and nutrient acquisition strategies. Practices based on promoting more sustainable ways of living through low-input and low-technology land-use practices that are socially acceptable should be supported to optimise plant growth and development, along with increased ecosystem diversity, productivity, and resilience.

Acknowledgements We are particularly grateful to the reviewers of each of the chapters for their role in improving their quality: Sayeda Abdelrazek Abdelhamid, Awang Azman Awang Pawi, Dibyendu Chatterjee, Andrew Dean, Clare Dean, David Elliott, Rosa Francaviglia, Krishna Giri,

Paul Grogan, Minakshi Grover, Catherine Hepp, Andrew Kugedera, Walter Leal Filho, Lo Chor Wai, Lungmuana, Newton Matandirotya, Ole Mertz, Gemma Muckle, Sharif Mukul, Arun Jyoti Nath, Alison Nord, Etsoshan Ovung, Dileep Kumar Pandey, Soubhagya Ranjan Padhi, Lh. Seitinthang, Giacomo Sellan, Robin Sen, Setiawati, Sileshi Gudeta Woldesemayat, Vibha Singhal, and Harold Vargas Hoyos.

References

1. Heinimann A, Mertz O, Frolking S, Christensen AE, Hurni K, Sedano F, Parsons Chini L, Sahajpal R, Hansen M, Hurtt G (2017) A global view of shifting cultivation: recent, current, and future extent. PLoS One 12:e0184479. https://doi.org/10.1371/journal.pone.0184479
2. Cairns MF (ed) (2015) Shifting cultivation and environmental change: indigenous people, agriculture and forest conservation. Routledge, London, UK
3. Kleinman PJA, Pimentel D, Bryant RB (1995) The ecological sustainability of slash-and-burn agriculture. Agric Ecosyst Environ 52:235–249. https://doi.org/10.1016/0167-8809(94)00531-I
4. Mertz O, Leisz SJ, Heinimann A, Rerkasem K, Thiha DW, Pham VC, Vu KC, Schmidt-Vogt D, Colfer CJP, Epprecht M, Padoch C, Potter L (2009) Who counts? Demography of swidden cultivators in Southeast Asia. Hum Ecol 37:281–289. https://doi.org/10.1007/s10745-009-9249-y
5. Nath AJ, Reang D, Sileshi GW (2022) The shifting cultivation juggernaut: an attribution problem. Glob Chall 6:2200051. https://doi.org/10.1002/gch2.202270081
6. Mishra A (2022) Shifting cultivation to sustainability—seeing beyond the smoke. Curr Sci 122:1129–1134
7. Padoch C, Pinedo-Vasquez M (2010) Saving slash-and-burn to save biodiversity. Biotropica 42:50–52. https://doi.org/10.1111/j.1744-7429.2010.00681.x
8. Finegan B, Nasi R (2004) The biodiversity and conservation potential of shifting cultivation landscapes. In: Schroth G, da Fonseca GAB, Harvey CA, Gascon C, Vasconcelos HL, Izac A-MN (eds) Agroforestry and biodiversity conservation in tropical landscapes. Island Press, Washington, DC, USA, pp 153–198
9. Pandey DK, Adhiguru P, Momin KC, Kumar P (2022) Agrobiodiversity and agroecological practices in 'jhumscape' of the Eastern Himalayas: don't throw the baby out with the bathwater. Biodivers Conserv 31:2349–2372. https://doi.org/10.1007/s10531-022-02440-7
10. Ramakrishnan PS (1992) Shifting agriculture and sustainable development: an interdisciplinary study from North-eastern India, UNESCO Man and the Biosphere Series, vol 10. UNESCO and Parthenon Publishing Group, Paris, France
11. FAO Staff (1957) Shifting cultivation. Unasylva 11:9–11
12. Tripathi SK, Vanlalfakawma DC, Lalnunmawia F (2017) Shifting cultivation on steep slopes of Mizoram, India: impact of policy reforms. In: Cairns M (ed) Shifting cultivation policies: balancing environmental and social sustainability. CABI, Wallingford, UK, pp 393–413. https://doi.org/10.1079/9781786391797.0393
13. Das D (2006) Demystifying the myth of shifting cultivation: agronomy in the north-east. Econ Polit Wkly 41:4912–4917
14. Ickowitz A (2011) Shifting cultivation and forest pressure in Cameroon. Agric Econ 42:207–220. https://doi.org/10.1111/j.1574-0862.2010.00507.x
15. Shimrah T, Rao KS, Saxena KG (2015) The shifting agricultural system (*jhum*) and strategies for sustainable agroecosystems in Northeast India. Agroecol Sustain Food Syst 39:1154–1171. https://doi.org/10.1080/21683565.2015.1089350
16. Wapongnungsang, Ovung EY, Upadhya KK, Tripathi SK (2021) Soil fertility and rice productivity in shifting cultivation: impact of fallow lengths and soil amendments in Lengpui, Mizoram, northeast India. Heliyon 7:e06834. https://doi.org/10.1016/j.heliyon.2021.e06834

17. van Noordwijk M, Hairiah K, Guntro B, Sugito Y, Ismunandar S (1996) Biological management of soil fertility for sustainable agriculture on acid upland soils. Agrivita 19:131–136
18. Changkija S. Festivals and ceremonies tie-up with shifting cultivation among the Ao-Nagas of Nagaland
19. Tripathi SK, Hauchhum R, Ovung EY, Singh NS, Vanlalfakawma DC, Upadhyay KK, Brearley FQ, Lalraminghlova H. Innovative shifting cultivation and other agricultural practices conducted by the Indigenous population of Mizoram, North-east India
20. Changkija S, Hombegowda HC, Manpoong C, Ray SK, Brearley FQ. Innovative Indigenous technologies for sustainable mountain farming systems in North-east India
21. Giri K, Mishra G, Jayaraj RSC, Tsopoe M. Agrobiodiversity and crop productivity of alder-based jhum farming system in Khonoma Village, Nagaland, India
22. Mishra G, Jangir A, Francaviglia R. Linking soil organic carbon and soil parameters in shifting cultivation sites of Nagaland (North-east India)
23. Hombegowda HC, Madhu M, Jakhar P, Vanitha SM, Raza MB. Recovery of soil organic carbon during forest fallow following shifting cultivation in Eastern Ghats and North-eastern hill region of India
24. Ovung EY, Brearley FQ, Singh NS, Tripathi SK. Differences in soil carbon and nitrogen across contrasting land-use systems in a sub-tropical hilly region, North-east India
25. Mishra G, Sarkar A. Ethnopedological knowledge of jhumias in Garo Hills, Meghalaya, to assess soil nutrients and health
26. Ray SK, Chatterjee D, Das SR, Ray S, Kandpal BK, Mishra VK. Agronomic aspects of nutrient management in shifting cultivation agro-ecosystems of North-east India
27. Lungmuana, Sangma CBK, Angami T, Shakuntala I. Shifting cultivation to settled agriculture for sustaining soil health and agriculture productivity in North-east India
28. Brearley FQ, Spencer M, Tripathi BM, Ovung EY, Singh NS, McGuire KL, Tripathi SK. Changes in bacterial community structure and nitrogen cycling gene abundance during the fallow recovery phase of shifting cultivation in North-east India
29. Momin MD, Tripathi SK. Screening of rhizosphere actinomycetes from shifting cultivation crops for pigment production
30. Mehetre GT, Deka P, Singh BP, Upadhyaya K. Sustaining shifting agricultural practices in North-east India through beneficial soil microbes
31. Debnath N, Laskar SY, Nath AJ, Pandey R, Das AK. Systematic review on shifting cultivation in North-east India: emerging research directions
32. Tripathi SK, Brearley FQ, Nath AJ. Management of soil fertility and crop productivity under shifting agriculture in Mizoram, North-east India: ways forward
33. Martin DA, Llopis JC, Raveloaritiana E, Coomes OT, Andriamihaja OR, Bruun TB, Heinimann A, Mertz O, Rakotonarivo OS, Zaehringer JG (2023) Drivers and consequences of archetypical shifting cultivation transitions. People Nat 5:529–541. https://doi.org/10.1002/pan3.10435
34. Kuotsuo R, Chatterjee D, Deka BC, Kumar R, Ao M, Vikramjeet K (2014) Shifting cultivation: an 'organic like' farming in Nagaland. Indian J Hill Farming 27:23–28
35. Mertz O, Bruun TB, Jepsen MR, Ryan CM, Zaehringer JG, Hinrup JS, Heinimann A (2021) Ecosystem service provision by secondary forests in shifting cultivation areas remains poorly understood. Hum Ecol 49:271–283. https://doi.org/10.1007/s10745-021-00236-x
36. Ramakrishnan PS (2015) Shifting cultivation and fallow management options: where do we stand? In: Cairns M (ed) Shifting cultivation and environmental change: indigenous people, agriculture and forest conservation. Routledge, London, UK, pp 186–198
37. Ziegler A, Fox JM, Webb EL, Padoch C, Leisz SJ, Cramb RA, Mertz O, Bruun TB, Vien TD (2011) Recognizing contemporary roles of swidden agriculture in transforming landscapes of Southeast Asia'. Conserv Biol 25:846–848. https://doi.org/10.1111/j.1523-1739.2011.01664.x

Festivals and Ceremonies Tie-Up with Shifting Cultivation Among the Ao-Nagas of Nagaland

Sapu Changkija

Abstract Every tribe has its own sense of historical identity which is reflected in oral traditions and beliefs that are passed down from one generation to another. The Ao-Naga tribe of North-east India has their own rich cultural heritage preserved in the form of oral traditions. The study upon which this chapter is based was carried out with the objective of detailing traditional ritual practices of sacrifices and ceremonies relating to the shifting cultivation (jhum) calendar system, which plays an important role in sustaining the tribe's cultural history. Documentation was gathered from oral narratives in the Changkikong range of Mokokchung district of Nagaland, from five individuals who had good knowledge of the belief system and ritual practices of the Ao-Nagas. The documented rituals are associated with festivals which are linked to their day-to-day life and are mostly performed for the welfare of the society with a particular focus on a bountiful harvest. With the need for such a harvest, religious ceremonies and rituals revolved around agriculture. The belief system in Ao-Naga practices is animistic in nature and polytheistic, worshipping several benevolent and malevolent spirits and gods, at different times and occasions for different purposes; they perform animal sacrifices by offering pigs, dogs, fowl, eggs etc., to pacify their spirits and gods. With the advent of Christianity, traditional celebrations are now performed with the Christian prayers during such events.

Keywords Culture · Religion · Rituals · Tradition

1 Introduction

The Ao-Naga are one of the major tribes inhabiting Mokokchung district of Nagaland, a state in North-east India, possessing rich traditional folk practices [1–3]. Nagaland has a total area of 16,579 km^2 and has a sub-tropical monsoon climate with a mean annual precipitation of 2000–2500 mm [4]. The indigenous inhabitants of the state

S. Changkija (✉)
Department of Genetics and Plant Breeding, Nagaland University, Medziphema, Nagaland 797106, India
e-mail: sapuchangkija@gmail.com

S. K. Tripathi and F. Q. Brearley (eds.), *Shifting Cultivation Systems*,
https://doi.org/10.1007/978-3-031-70388-1_2

have racial and socio-cultural affinities with South-east Asia and comprise 16 tribes recognised by the Indian Government. Nagas have a relatively high literacy rate mainly as a result of the coming of American missionaries to the state and their conversion to Christianity [5]. Prior to the advent of Christianity, they were animist with the belief in supreme creators called 'tsungrem'. The name 'Ao' is derived from 'Aor' meaning 'those who went' (i.e. across the Dikhu river between Tuensang and Mokokchung districts of Nagaland [1, 2]). The Ao are one of the largest tribes of Nagaland, and two different dialects are spoken by them, the Chongli and the Mongsen dialects. Among Ao-Naga it is said the kinship has evolved from six stones representing six major clans [1]. Some of the distinctive characteristics of the Ao are the custom of disposing dead bodies on raised platforms (this custom is now abolished), tattooing of the womenfolk, use of huge xylophones hewn out from a single log, elaborately organised village councils, and smoking tobacco through pipes by both men and women [2, 3]. Like most of the Indigenous tribal communities in world, the Ao-Naga believe in the existence of supernatural powers. In Ao vocabulary, the term tsungrem is used to denote several gods such as Lijaba as the creator of the universe, Meyutsungba as the keeper of the land of the death, and Aningtsungba or Longtitsungba as the lord of the sky or heaven. Besides these gods, there are also gods associated with objects of nature such as mountains, big stones, rivers, forests etc. [6].

Shifting cultivation is a form of low-input agriculture and fallow management that is common in North-east India, particularly for the cultivation of paddy (*Oryza sativa*; Poacaee), casava (*Manihot esculenta*; Euphorbiaceae) maize (*Zea mays*; Poaceae), taro (*Colocasia esculenta*; Araceae), cucurbits and legumes [7, 8]. It is also commonly referred to as swidden cultivation and locally in North-east India is called jhum cultivation. In this system, the forest is slashed-and-burned followed by growing of various vegetables and food crops using minimum tillage practices and maintained with rotation and sequential cropping systems. These help in weed control and enable crops to grow with reduced competition. Shifting cultivation is a sustainable practice in sparsely populated areas if a long duration fallow is maintained [7, 9]. It is the most widespread type of cropping system and is directly supported by the forest ecosystem. Basically, shifting cultivation system is a reflection of human–environment relationships in this sub-tropical mountain region which evolved along with suitable forest management. This system is characterised by a highly diversified cropping pattern meeting the peoples' needs for food, spices, medicines and aromatic plants, while maintaining the rich germplasm of the indigenous crop varieties. It has been in use for centuries and still remains a major land-use practice, which provides a basis for subsistence farming, maintenance of cultural values and social stability for the people living at low population densities.

In the tribal mountain communities of the region, shifting cultivation is a way of life and all elements of their social life are geared to, and integrated into, the continued operation of their culture, festivals or festivities that are closely associated with the annual jhum calendar (Table 1) as well as worship to gods and making offerings from the jhum, and this is one of the important reasons why people are resistant to any suggestions regarding cessation of jhum cultivation. Rituals, rites, prayers, sacrifices and offerings hold true in the religious life of most of the tribal communities. Religion

Table 1 Calendar showing the sequence of operations during jhum cultivation in the Ao-Naga practice

Month(s)	Shifting cultivation operation(s)
Dec–Jan	Survey and location of household fallows and slashing of jhum fields
Feb–Mar	Burning of the plot followed by construction of field house and field preparation
Mar–Apr	Sowing various early seeds and broadcasting paddy seeds
Apr–May	Germination of seeds and succession of early weeds
May–Jun	First weeding after pre-monsoon rain
Jun–Jul	Start of the harvest of some early crops
Jul–Aug	Second weeding whilst continuing crop harvest
Aug–Sept	Third weeding and continuation of crop harvest followed by protection of crops against depredation by wild animals, pests, etc.
Sept–Oct	Harvest of paddy (main crop)
Nov–Dec	Storing paddy followed by harvest of legumes and cucurbits
Jan–Mar (following year)	Harvest tuber crops (early part of next year)

is defined as any set of attitudes, beliefs and practices pertaining to supernatural powers, whether that to be forces, gods, spirits, ghosts or demons. The Ao-Naga traditionally observe a series of ritual ceremonies at every important phase of the jhum calendar. These ceremonies are observed in order to invoke the blessing of the supreme god to ensure good crop health and good harvest, protection from pests, wild animals and natural calamities, and also for good health. There are many ways of propitiating the spirits: one common form is to offer food items such as eggs, meat, rice or rice-beer. There is great variation in the details of rituals among Naga tribes or even from village to village. There are specific persons who perform these rites; they are sometimes village priests called 'patir/putir', and also the herbalists or medicine men called 'arasentsur'. Worship of the gods follows an agricultural cycle because of the belief that they control the elements of nature. Propitiating them is important for the well-being and prosperity of the people. In general, a ceremony is called 'among' which is of two categories (a) a period of purification and taboo called 'anempong' and (b) performance of sanctity with restrictions called 'kimoh'. While in anempong, abstention from sexual activity and certain foods is observed, only normal household activities are conducted, and during this period some particular plant species (e.g. *Artemisia nilagirica* (Asteraceae), *Litsea cubeba* var. *cubeba* (Lauraceae), *Ocimum americanum* (Lamiaceae) etc.) are kept inserted through the bamboo wall in the entrance of the house signifying the sacred performance. In kimoh, entry to one's home is restricted by keeping the entrance door in a horizontal position in the daytime (doors are usually made into an interwoven flap) whilst also abstaining from sexual activity. Thus anempong and kimoh are necessary purification processes, which are observed before the actual ceremony, among, and are continued till the final rites are over.

2 Materials and Methods

This documentation is mainly gathered from oral narratives in the Changkikong range of Mokokchung district of Nagaland during the year 1998–2002. Oral narratives were collected exclusively from five individuals (names of the informers are listed in the acknowledgments) who had good knowledge on ritual practices and belief systems of the Ao-Nagas. Secondary source material, in the form of published or unpublished written material, books, articles, and documentation by folklorists available in the form of published anthologies, was also consulted to support data collection.

Glossary

Aam/Amchi: leaves of *Musa* (Musaceae) or *Phrynium* (Marantaceae) species

Akhi: conical-shaped field basket

Aling: bamboo splits/rope

Aluthi/Aluthima: jhum field house

Aluthini: location of the jhum field house

Among: ceremony/ies; also means a sub-ceremonial altar

Amtsang: cup or plate made from leaves, e.g. of *Phrynium* (Marantaceae) species

Anempong: a period of purification through abstention of sex, certain food and other activities

Ankhang: cage for fowl

Aokhang: pig cage or pigsty

Apen: Cylindrical wooden drum made from cow hide

Aphen: second-year jhum field

Aphu: y-shaped pole/stick in the phumendang

Arasentsur: herbalist/medicine man

Ariju: morung/men's institute

Arren: attraction power associated with the spirit would

Arrenlong: round sacred fertility stones

Asemni among: three days ceremony/sabbath

Ayeibenlu/Azubenlu: dog sacrifice field location

Ayeilepok/Azulepok: dog sacrifice

Chubangchanger: members of the morung/men's institute

Chubangmenti: morung/men's institute centre

Dao: large multi-purpose knife similar to a machete

Gaon bora: chief of the village

Henmi: large intestine

Henmitsuba: fowl sacrifice

Hentsukhang: egg basket

Jenti: main celebration day

Khang: cage/basket
Khel: colony of clans
Kimak: household visits
Kimoh: family ritual of sanctity with abstention of sex, aversion to meat and fish, and prevention of others entering the family house
Kulemtong: worship tree, usually a large tree in the sacred forest near the village
Langtem: loin cloth
Lijaba: creator god
Ludhi: first-year jhum field
Mangkotsurong: location on land belonging to a hereditary representative of former chiefs in which head trophies are kept and oath ceremonies are conducted
Mejemtzu: rice-beer
Meken-kum: year of reduced rainfall
Mepuzen: eldest member of the sector
Mesen: insects
Metsu: altar pole
Mezen-kum: year of normal rainfall
Milling: fire thong
Moatsu chilusa: dancing procession
Molotsu: ocean
Morung: institute/dormitory of men in the villages
Mungti phumendang: main village community altar site with stone monolith
Nahsem: sharp bamboo blade made of *Pseudostachyum polymorphum* (Poaceae)
Nisungsangla: humans
Nokleptsu/Nokuza: dao, scabbard and belt
Oktso: sharpened bamboo with a pointed tip
Ookla: male pig
Patir/Putir: village priest
Phaba: to catch
Phuchung: altar erection ceremony
Phulep: cutting of altar pole
Phumendang: altar
Pouk: jhum plot
Sungpunu: firewood collection day
Sungpur: youngest of three age groups among the men
Takzu: bamboo cup
Tenem mahlu: hornbill play
Tsadem: lunch packed within palm leaves
Tsaktangma: stage of paddy before the formation of grains
Tsungli: ceremony offering food to the creator god Lijaba

Tsunglushi: rain shields made from palm leaves
Tsungrem: collective name for the gods
Tzu-ok: ceremonial community fishing
Tzupa: springs/ponds
Yangar: community institutional age groups
Yimgkong kulem/Yimgkong phumendang: village altar

3 Some Traditional Folk Practices and Beliefs of the Ao-Naga

3.1 Management of Tree Species

In the jhum cultivation system, the forest is slashed and burnt but, during the process, the big trees are traditionally merely lopped. In the villages, the rich men leave a few branches uncut at the top of the tree. This is apparently an idea of leaving a place of refuge for jungle spirits. The practice is regarded as merely an indication of wealth and it is believed that, should a poor villager show such an impertinent spirit as to imitate his rich neighbours and leave a few branches uncut, so he will also have a good crop. The top branches of big trees such as *Alnus nepalensis* (Betulaceae), *Betula alnoides* (Betulaceae), *Ficus benghalensis*, *Ficus elastica* (Moraceae), *Schima wallichii* (Theaceae), *Stereospermum chelonoides* (Bignoniaceae) and in the genera *Caryota* (Arecaceae), *Castanopsis* (Fagaceae), *Eurya* (Pentaphylacaceae), *Quercus* (Fagaceae), *Terminalia* (Combretaceae) etc. are left uncut. The idea as it exists may be the belief of an association between the trees as the abode of jungle spirit and prosperity. Ao-Naga believe that the 'arren' (attraction power) associated with the spirit would certainly be attracted to the fields and to the lopped trees of the rich man as 'arren always attract arren'. Apart from slashing of the jhum, the Ao-Naga community does not cut trees or wood randomly throughout the seasons for domestic uses. The constructional wood and bamboo items are normally obtained during the last part of autumn and winter in a system which protects infestation by various insects and from wood borers. Fire wood is also only harvested/cut during early spring coinciding with the timing of slashing of jhum. Cutting of trees is done following the phases of the moon and occurrence of different cicada sounds.

3.2 Oath-Taking Tree: Mangkotsurong

Erythrina arborescens and *Erythrina stricta* (both Fabaceae) are called 'tsubentong'. These trees are planted outside the 'morung' (institute/dormitory of the men) in every Ao-Naga village, and surrounded by round sacred stones called 'arrenlong' (fertility stones). The place where this tree is planted is called ‘mangkotsurong’ in which the oath ceremony is taken, Earlier, during the head-hunting days, the enemies’ heads, hands and feet which were brought as war trophies were kept here hanging from these trees after performing the various rites and ceremonies connected with head-taking. These trophies were kept on display for two to three months, hanging from the tree, after which they were moved and kept in the respective entrances of the warrior households. The oath taking during misunderstanding within the community is done in this mangkotsurong or under the *Erythrina* tree which is grown on land belonging to a hereditary representative of former chiefs. Oaths are taken by pulling the leaves which, it is believed, do not come away in the hand of a false swearer. As the leaves are pulled the oath-taker swears that if he is wrong he/she will fall and would rot like the *Erythrina* tree (this species, when fallen, decomposes rapidly) and that he/she will also rot like the enemy heads on the tree which are kept on display and seen by all the villagers. The wood of this tree is not used as fuel during marriage feasts. At the time of peace-making between enemy villages, each party would bring one of the arrenlong from the other’s mangkotsurong and place it under their respective trees, knowing that till the stone rots there will be no more war or head-taking between them.

3.3 *Sacrifices and Offerings*

Every ritual ceremony is observed with ceremonial sacrifices of eggs or a chicken or a pig or a dog depending on the type of ceremony, along with rice-beer called ‘mejemtzu’ and ginger. In public/communal ceremonies, the sacrifices are made at the village alter called ‘yimgkong phumendang’ located on the wayside of the field by the priest who is the eldest of the village or any elderly male sibling. In case of individual ceremonies, sacrifices are made at their individual altar (‘phumendang’) in front of their field house (‘aluthi/aluthima’) by the head of the household. The words of prayer during the sacrifices may differ from ceremony to ceremony. The livestock, eggs, ginger, rice-beer and the leaves called ‘aam’ (*Phrynium pubinerve*; Marantaceae) or banana (*Musa* sp.; Musaceae) are always kept ready the night prior to the selected day of the ceremony and are carried in their respective cage/basket (‘khang’) specially made for it such as ‘aokhang’ (pig cage), ‘ankhang’ (fowl cage) and ‘hentsukhang’ (egg basket).

3.3.1 Tsungli

Tsungli ceremony is the offering of food to the creator god Lijaba. It is done at the phumendang where sacrifices are made. The ceremony consists of offering cooked rice, ginger (*Zingiber officinale*; Zingiberaceae), common salt, dried local fish, rice-wine or rice-beer (mejemtzu), eggs, and liver of fowl or pig, depending on the type of sacrifices made on the day. When the cooking of the food is completed, the man takes an 'amchi' leaf (leaf of *Musa* sp. or *Phrynium* sp.) and shapes it into palm size strips of plates called 'amtsang'. Six amtsang plates of such strips of amchi leaves are laid in a row in the phumendang and small pieces of all the food items are served. One round of an item followed by another item from all the plates is made naming each item saying "*lo-o acha*" (offering rice), "*lo-o sungmok*" (offering ginger), "*lo-o ngatsu*" (offering dried fish) etc. In naming the items, the source is also mentioned i.e. rice and rice-beer made from a particular field, fish from a particular river: such as Chubi, Menung, Milak, Tzula, Tzurang etc. Having done this, six small pieces of amtsang are prepared from the amchi leaf and those are grouped into two sets and placed on both sides of the phumendang. Rice-beer is poured out of the bamboo cup called takzu which is about 8 cm diameter and 25 cm deep. After completion of all the necessary rituals of the tsungli offering, the remaining portion of the food is eaten by the priest, their family members and all who attended the ceremony.

3.3.2 Pig Sacrifice

The pig sacrifice is done by a priest with a prayer. The pig is taken out of the aokhang (pig cage), both the front and hind trotters are bound by 'aling' (bamboo splits/rope). A pole of bamboo or wood, about 8 cm in diameter and 1 m in length is inserted between the right and left fore and hind legs and the pig is placed in the phumendang on the left side. The man takes an 'oktso' (sharpened bamboo with a pointed tip) and gives a marking in each point of prayer from shoulder through the side to the rump of the pig. After the prayer, the pig is killed by piercing it just below the right side of the fore limb so to pierce the lungs and heart for a rapid death. Later, it is dressed and cooked for the meal. A piece of the liver is also kept aside to offer to Lijaba along with other ingredients for the tsungli offering.

3.3.3 Fowl Sacrifice

Fowl sacrifice is called henmitsuba in which a healthy and un-warded fowl is sacrificed. The fowl is taken out of the 'ankhang' (fowl cage) in the phumendang and both wings are held together by the left hand and a set of feathers is plucked from the saddle-back during the prayer and placed in the phumendang. After the words of prayer are completed, the fowl is killed by slitting the throat with a sharp splint blade of a specific bamboo species (*Pseudostachyum polymorphum*; Poaceae) called 'nahsem' and the blood is drained out through the cut and also smeared on to the ankhang with the help of another man. After that, the abdomen of the fowl is cut open, and the entire large intestine ('henmi') is pulled out along with the cloaca to be left fastened along with the plucked feathers of the wings in the ankhang. A piece of the liver is also kept aside to offer to Lijaba along with other ingredients for the tsungli offering. The omen is taken by observing the right and left side of a leaf of *Phrynium pubinerve* by tearing. The omen performance may be for good health, good harvest of the crops, to protect against weed and pest infestation in the field, for prosperity in life etc. This decorated ankhang is placed horizontally in the extension bamboo of the 'aphu' (y-shaped pole/stick) in the altar as the share for Lijaba. The fowl is then dressed and the meat is cooked for the family to eat.

3.3.4 Egg Sacrifice

In the egg sacrifice, the egg is taken out from the hentsukhang (egg basket) and sacrificed with prayers by holding it in the right hand and cracking it slowly on a small stone provided for the purpose in the phumendang. The yolk and albumen is poured in an amtsang (a conical cup made of banana or *Phrynium* leaf) and cooked upon a fire till it is coagulated. The eggshell is put inside the hentsukhang again and fastened horizontally in the aphu (y-shaped pole/stick) in the phumendang.

4 Sequence of Ceremonies Associated with Shifting Cultivation

The regular ceremonies that are related to shifting cultivation are observed by the Ao-Naga in their annual cycle of agricultural activities followed during annual cropping. In certain situations, some incidental or emergency ceremonies may occur or sacrifices may be performed whenever circumstances such as accidents in the field or natural calamities like flood, drought, insect and pest infestation of crops arise.

4.1 Regular Ceremonies

4.1.1 Sungkhomesa

A one-day ceremony is celebrated before a new jhum plot ('pouk') is cleared. In this, the whole village observes anempong (a day of sanctity with abstention of sexual activity and certain foods). The village priest takes a red cock and sacrifices it with prayers followed by a tsungli offering in the proposed plot appealing for protection from the evil eye, injuries and death during forest clearance (note that nowadays, a prayer day is observed in every Ao-Naga village initiated by the pastor or the church leaders for the same purpose). Slashing of forest for new jhum fields only takes place after this ceremony.

4.1.2 Talen/Luen Walok

When the villagers decides to shift the fallow to the new area for the year, the priest sacrifices an egg and a fowl to Lijaba on the path leading to that fallow area. A tsungli offering is observed to sanctify the way to ensure safe passage to-and-fro. It is also observed after burning the slash. The Talen/Luen walok ceremony is also performed after any accidental occurrence in the jhum field or in the village to drive out the evil spirits and the family performs kimoh (it is observed for a week where the family performs sanctity with abstention of sexual activity, aversion to meat and fish and preventing others from entering the family house).

4.1.3 Luzha and Sungkho Lepden

After the sungkhomesa and the talen/luen walok ceremonies, the head of the family goes to the proposed field plot and the area is surveyed and inspected for locating the 'aluthini' (field house location). Then a small space is cleared and he takes the omen with a 'milling' (fire thong). Here dry 'tangmu' wood (*Rhus chinensis*: Anacardiaceae) is placed on the ground and he then presses his foot on the wood. A finely dressed split bamboo and dry leaves and twigs are placed under the wood. He then grips either end of the split bamboo, pulls it rapidly in a right to left, or vice versa, manner until the split bamboo snaps. The omen is then observed by studying the frayed fibers of the bamboo. Another way of observing the omen is by performing 'phulep' (cutting of the altar). Here, an aphu/phumendang is set up by erecting a post called 'metsu' (altar pole), usually a pole of 'alumeset' (*Eurya acuminata*: Pentaphylacaceae), 'wazamama' (*Ficus* spp. Moraceae) or 'sungken' (*Macaranga indica*: Euphorbiaceae) about 5 cm in diameter and 2 m long. Six cuts are made by a 'dao' (knife similar to a machete) on the post at intervals of approximately 10 cm. As every second cut is made on the post, a splinter of wood is carved out of the post. The omen is predicted according to which side the inner side of the splinter falls on the ground. Then the head of the family notes his dream that night; if the dream indicates that all will be well he goes alone or with his wife to the field, clears the aluthini (location of the field house) and sacrifices an egg by cracking it on a small stone in the phumendang with a prayer that he may have good harvest and be safe from sickness of his father and forefathers. The yolk and the albumen of the egg is poured in an amtsang (banana or palm leaf cup) and cooked upon the fire. A tsungli offering is made in the phumendang. The egg basket with the egg shell is tied in the aphu/phumendang and the egg is eaten. The farmer then starts the forest clearance. Since almost all the Ao-Nagas have now been converted to Christianity, the farmers now begin with a Christian prayer and dedication for blessing and good health throughout the year.

4.1.4 Merokmesa

After the field is burnt, the village priest and the representative of each of the clans in the village go to the 'yimkong kulem' (village altar) the next day and sacrifice an 'ookla' (a male pig) on behalf of the whole community. The purpose was that if any wildlife has been burnt alive during the burning of the forest vegetation, it was believed that the weeds would overgrow the cultivated plants and would result a poor harvest. Therefore, this ceremony is observed in order to avert such a situation. A ceremony is observed and following the tsungli offering and the remaining pork is consumed by the priest and the people who accompanied him.

4.1.5 Alumesa

The first day a man goes to the field after the merokmesa ceremony, he carries an egg to make a sacrifice. A site is selected to set up the aphu/phumendang (altar) in front of the aluthini (field house location). Then a bamboo or wooden post of 5–10 cm diameter is erected and the egg is sacrificed by cracking it slowly on a small stone near the post with a prayer that the 'spirits of any animals or snakes burnt in the field may be rendered powerless and that bumper crops and good health may be granted'. The yolk and the albumen of the egg is poured in an amtsang (cup made of banana or palm leaves) and cooked upon the fire. Tsungli, which is the offering ceremony, is done near the altar. The remains of the egg are eaten by the family members in the field. After this they start collecting materials for the construction of the field house. Sometimes a dog is sacrificed (known as 'ayeilepok/azulepok') and the location is known as 'ayeibenlu/azubenlu'.

4.1.6 Phuchung

Phuchung is the ceremony observed during the erection of the altar in the field which may be done before or after the sowing of the paddy seeds. It is done both in the 'ludhi' (first-year field) and the 'aphen' (second-year field), although the work in the aphen is just repairing and extending the old altar. In the ludhi, the new altar and place for offerings is constructed in front of the aluthi/aluthima (field house) where the family offers sacrifices. On that day the whole family goes to the field taking with them a small pig, a fowl and an egg. The ceremony is done through the tsungli offering. All the sacrificial items are carried separately in their own designed cages/baskets, i.e. aokhang (pig cage), ankhang (fowl cage) and hentsukhang (egg basket). On reaching the field, the husband collects six bamboos of about 3 cm in diameter and 2 m in length along with some extra for other activities. The phumendang is constructed by setting up the six bamboos in a criss-cross manner in front of the aluthi/aluthima. All the sacrificial works and the offerings are made in this phumendang throughout the year. The construction of the bamboo structure in a criss-cross manner is for fastening the baskets in which the sacrificial items are to be brought to the field throughout the year in various ceremonies. A prominent monolith (stone) which is kept in the vicinity of the village since the time of the village foundation is called 'mungti phumendang' (main village community altar site with a stone monolith). On occasions the among (a sub-ceremonial altar) is moved to another place out of the village or to the community field and to the individual fields if a sacrifice has to be undertaken. In general, after the construction of the phumendang is over, the tsungli offering is led by the priest or the man of the household. Firstly, they sacrifice an egg with the prayer to Lijaba. Then the yolk and the albumen is poured in an amtsang (banana or palm leaf cup) and cooked upon the fire. Secondly, a fowl is sacrificed in the phumendang with the same prayer. The omen may be for health of the family, good harvest, preventing weed infestation in the field or prosperity in life. The henmi (large intestine) and some feathers are fastened in the ankhang (fowl cage), smeared with the blood of the fowl taken from the cut area in its neck and fastened to the altar. Then the fowl is dressed and cooked and consumed by the family. The family pig is also sacrificed at the phumendang following the pig sacrifice process by a priest with a prayer. The empty cage of the pig is fastened to the altar. This is followed by the offering ceremony tsungli. The remaining portion of the sacrificial items are eaten by the priest and family after which they return home in the evening. The family then observe anempong for six days.

4.1.7 Matsu Walok

The sowing of the seeds in the aphen (second-year field) starts with a ceremony called matsu walok (sanctification of seeds). In this ceremony, two village priests collect seeds of every variety of crop from the village rich men's house and sanctify the seeds in the phumendang near the village field path just outside the village gate by sowing a few sample seeds. A pig is sacrificed and the tsungli offering is done. After this, the priests and the village elders feast on it and the family observes kimoh for six days. The plot where the seeds are sown is carefully fenced on the same day. In the present day, the church organises a day of prayer for all the villagers in the church to sanctify the seeds before sowing. This ceremony is performed to ensure successful germination, safeguard against insects, pest and for good harvest.

4.1.8 Tenden Mong and Matsu Palak

When the first-year jhum field is ready for sowing of paddy seeds, this ceremony is observed with anempong. On the appointed day, all the villagers stay in the village. A village priest goes to the yimgkong phumendang (village community altar) with a fowl and some paddy seeds. A little space is cleared and the seeds are sown and fenced. Then he takes the fowl, holding both the wings with the left hand and utters the words of prayer to Lijaba. A set of feathers are plucked out by the right hand from the back saddle of the fowl in each point of the prayer. He then requests Lijaba for good seed germination and protection from animals so that 'nisungsangla' (human beings) can live a happy life next year too. Then the fowl is killed by slitting the throat with a nasehm (bamboo blade made of *Pseudostachyum polymorphum*; Poaceae). The omen is then taken by pulling out the large intestine (henmi). The fowl is dressed, cooked and a tsungli offering is completed. The remaining portion is eaten by the priest. He then returns home and observes anempong. Following the tenden mong, every household can go to their own field with an egg for matsu palak (starting of seed sowing). The egg sacrifice is performed and after the offering is made, sowing of the seeds starts from the site of the altar and then extending to the rest of the field.

4.1.9 Moatsu

Moatsu is the main festival of the Ao-Nagas. This festival is celebrated during the spring, during which people are praying for germination and good growth of the crops and blessings throughout the year. It is celebrated with joy and great enthusiasm immediately after sowing of the seeds in the ludhi (first-year jhum field). The moatsu celebration lasts for five days. As the sowing time differs from place to place so the date of moatsu festival also differs. Longsa village is the first to celebrate this festival followed by other villages. But nowadays this festival is celebrated simultaneously on a fixed date i.e. from 1st to 6th May in every Ao-Naga village or wherever they reside. The first day is called 'sungpennu'. On this day the 'sungpur' (youngest of the age groups among the men) in the 'ariju' (morung or institute for men in the village) collect the reeds, firewood, water and other materials which will be required during the festival. They prepare a drum called 'apen' made of a wooden cylinder with cowhide and cane in the ariju On the same day every 'chubangchanger' (members of the morung), formed of all three age groups, collect firewood, water etc. in the 'chubangmenti' (morung/men's institute centre). The second day is the 'okpani' (the day of pig killing). On this day every ariju and chubangchanger kills a pig. Every household contributes scented sticky/glutinous rice, rice-wine, rice-beer etc. for the festival. All males in the village cut their hair on the second day. In every chubangmenti, extension or repairing of sitting places are also done. The third day is the 'jentini'. It is a day of anempong during which pounding of rice, carrying of water from the well, washing of cloths, sexual activities etc. are restricted. All the chubangchanger gather in the chubangmenti and they spend their day singing, eating, drinking and merrymaking. From the fourth day onward, every family opens a new wine pot. Attired in rich costumes and fineries according to their status, the chubangchanger visits every family in groups. This visitation of houses is called 'kimak'. They sing songs of praises at each house they visit until the next morning. The next two days are the days for the 'moatsu chilusa' (dancing procession) where all the men and women put on their traditional costumes and dance. The dancing party from the lower elevation part of the village goes to the upper part of the village whilst the party from the upper part heads downwards. The dancing is led by the apen (drum) and the dancers dance to this drum. At the end of the procession, four people perform the cock-fight dance while the rest of the people on both sides witness it. The dancing party from the lower part of the village is given reception at a house in the upper part, and vice versa, where rice-beer and food are provided. They visit the homes of friends and sing songs of glory and honour. On the sixth day, which is the last day of the festival, 'tenem mahlu' (hornbill play) is performed, and a tug-of-war is conducted between the men folk and the women folk. For this purpose, the vine of 'kheling ninang' (*Entada rheedii*; Fabaceae) is used. The particular vine is collected by the sungpur (youngest of the age groups among the men) on the first day of the festival. In the evening, the same activities carried out on the fifth day are repeated but on the way back to their respective sector, a particular dance called 'tenem mahlu' (hornbill play) is performed. Everyone in the village eagerly waits until midnight to watch the tenem mahlu dance as the dancing party returns to their respective sector only at midnight. When the dancing comes to an end they usually sing a song as a mark of completion of the moatsu ceremony and this also serves as an announcement of activities that will follow after moatsu.

4.1.10 Morepden Mong

Morepden mong, meaning weeding ceremony, is observed every year just before the beginning of the weeding season and after the moatsu festival is over (in the month of May). On the appointed day, the villagers stay in the village observing anempong while the priest goes to the place where he sowed the seeds during the tenden mong sacrifice. Some of the young leaves are plucked and chewed uttering words like "*oh, these leaves taste bitter, they taste sour* etc." It was believed that those words would be heard by Lijaba and thus would prevent insects and other pests like rodents, grasshoppers etc. from destroying the tender leaves. The priest sacrifices a fowl and an egg with a prayer which is followed by a tsungli offering. After this ceremony, every family carries an egg to their field and sacrifices it by cracking it in the phumendang. After this offering is made along with rice, ginger, local dried fish, egg and rice-beer in the phumendang. The family then observes anempong for one day.

4.1.11 Tsadem Mong

When the paddy ripens, the priest who performed the morepden mong goes to the field where he had sown the seeds during tenden mong near the same altar. He takes a fowl and an egg with him and sacrifices it to Lijaba with a prayer. After the prayer the fowl is killed, the omen is observed, and the chicken is dressed and cooked. The egg is cracked open on a stone and the yolk and the albumen is poured into a palm leaf cup and cooked upon the fire. Then a tsungli offering is done with rice, ginger, dried fish, egg and the liver of the fowl in six plates of amchi leaves. The cage/basket of the fowl and the egg are fastened in the altar. The priest reaps a few panicles of the ripened paddy which is then threshed and carried home. The paddy is dried and husked the same evening. Then the rice is cooked and eaten chanting these words: "*this rice taste bitter, this rice taste sour* etc." In this way the priest offers thanksgiving to Lijaba for the good crop. He uses the words "*bitter*" and "*sour*" in order to deceive the pests like the rodents so that they do not eat or destroy the grains while it is stored in the granary. The next day onward all the family members from every household go to the field with a fowl and an egg as 'tsadem' (packed lunch). In the morning before going to the field, three small packets containing cooked rice, local dried fish, ginger and a sacred herb called 'aho channi' (*Hypericum japonicum*: Hypericaceae) with soil still on its roots are tied in the husband's 'akhi' (conical-shaped field basket). The egg is offered to Lijaba on the altar with a prayer. During the prayer, the egg is slowly cracked open on the sacrificial stone in the altar. The yolk and the albumen are poured into a cup of palm leaves kept ready for the purpose and then cooked. The fowl is also taken out of the basket and sacrificed on the altar with the same prayer used during the egg sacrifice. Here both the wings of the fowl are held by the left hand and the words of prayer are offered. The fowl is then dressed and cooked. Tsungli is performed in the same way as in the 'phuchung' (altar erection ceremony) with rice, ginger, local dried fish, small pieces of egg, liver of fowl in six plates of amchi leaves in which rice is served and in six cups made of the same amchi leaves. The same food package preparation called amtsang, along with the sacred herb of aho channi, is made into small packets and bound with paddy stalks in the phumendang. When the tsungli is over, the harvesting commences. The panicles of the paddy are cut using a sickle, it is then put in the akhi (conical-shaped field basket). which is carried on one's back while it is filled with panicles and the paddy is then collected in the field house and threshed. The paddy is separated from the old stems and winnowed to remove the grain that is boiled slightly, dried in the sun and carried home in the afternoon. The dried rice is husked and cooked the same evening. No one is allowed to eat the rice grains during husking or cooking. The pot in which the rice is to be cooked is adorned with packets of the tsungli food items like rice, ginger, local dried fish and aho channi by binding it in the neck of the pot, and the rice is then cooked. Family members enjoy the first harvested rice of the year with thanksgiving to Lijaba. Tsadem mong is celebrated as thanksgiving to Lijaba, as well as for lasting harvest in the field. Nowadays, every year, a thanksgiving day is organised by the local churches. On this day all the best products of the crops and handicrafts as well as cash are collected in the church and thanksgiving prayers are offered in the Christian way. The proceeds of these materials are used for missionary works.

4.1.12 Tsungrem Mong

Tsungrem mong (sabbath) is one of the most important ceremonies, or among, observed by the Ao-Nagas. It is observed for three days so it is also called 'asemni among' (three day sabbath). A day of worship and sacrifice offered to Lijaba for a plentiful harvest of crops and good health. It is the most strict and holiest of ceremonies among all the Ao-Naga ceremonies. Failure to observe anempong is considered to result in drought, hailstorms, landslides, floods and other natural calamities that destroy the crops or cause illness, epidemics etc. The tsungrem mong celebration is observed in the tsaktangma (stage of paddy before the formation of grains and harvest). When the panicles are at this stage, the ceremony is celebrated by praying to Lijaba for a bountiful harvest. In the earlier days, the Ao-Naga villagers celebrated this festival in their own convenient days. Accordingly, Longsa village observed the tsungrem mong festival earliest followed by other villages in the Ongpangkong range. Nowadays, tsungrem mong is observed from the 1st to 3rd of August every year in every village and town.

The important events during the tsungrem mong are the exchange of gifts like new tobacco pipes made of bamboo from the boys to the girls, and the girls give 'nokleptsu/nokuza' (dao, scabbard and belt) to their boyfriends. The feast of the yangar (community institutional age groups) in one of the girls' houses with merry-making and singing is another event. There is a formal announcement of wedding engagements to be solemnised after the harvest season by accepting meat sharing during the festival. The first day of the tsungrem mong is the 'sungpunu' (firewood collection day) where collection of firewood, water etc. takes place. The wine which has been kept ready in advance in every house for the festival is checked. A pig is sacrificed outside the house of the 'mepuzen' (eldest member of the sector). Three to four small pieces of pork are fixed in a small bamboo stick and distributed from house to house. These strips of meat are stuck on the wall near the door of every house. The family collects the pork pieces after sunset, roasts them upon the fire and distributes them among the children. The whole family observes anempong from that day onward. The newly constructed houses meant for the couples expected to marry after the harvest are also given the shared meat. If they accept the meat it signifies that that they are getting married after the harvest. The second day is the 'jenti' (main celebration day) which is a very strict among (ceremony) as no-one may leave the village nor husk rice or fetch water from the well; anempong is observed and rice-beer is offered to the visitors. The third day is a less strict day where restrictions on activities like fetching water from the well etc. may be relaxed. In some villages, traditional games and tug-of-war between the male and female members of the respective age groups take place. In the evening, the family members remain in their houses. Everyone shuts their door before dusk, sits for the evening meal and silently stays indoors. On that day the door of every house is adorned with 'angpetwa' (leaves of *Zanthoxylum* sp.; Rutaceae).

4.1.13 Phukup

This ceremony is thanksgiving celebrated after the harvest of the paddy and other crops in the aphen (second-year jhum field). It is a ceremony to bid farewell to the old jhum field and an invitation to Lijaba to come in search of the next year's plot. In the ludhi (first-year jhum field), phukup is celebrated as thanksgiving as well as prayer to Lijaba for the blessing on the day of paddy threshing. For this ceremony, an auspicious day is fixed and the whole family goes to the field carrying a small pig, a fowl and an egg. The method of sacrifice performance in the phukup is similar to the phuchung (altar erection) ceremony and is celebrated with feasting. In the evening, they leave the field with all the implements and vegetables. But just before leaving they collect the red cockscomb flower (*Amaranth* sp.; Amarathaceae) and decorate the altar with it so that the field deity may think that the nisungsangla (humans) are still standing in the aluthi/aluthima (field house). It is a traditional practice to plant the cockscomb surrounding the field house and the 'tzupa' (springs/ponds) to please the field deity.

4.1.14 Mera Mija/Mera Nuja and Kimang Mija/Kimang Nuja

This harvest ceremony is normally performed by an affluent village man who invites the whole village or 'khel' (colony of clans) or by the host's clan in order to achieve the honour. The affluent farmer normally cultivates a large jhum field and the ceremony is performed as a legacy of his hard work and display of wealth, which brings honour and respect. The harvest should be done on that day itself by the whole village or by the khel or by the clan community. This task was meant to challenge the upcoming generation to inculcate and uphold the rich heritage of their forefathers and to be hard-working farmers. While hosting mera mija/mera nuja all the villagers of the khel or the clan community are invited to partake in the harvest along with the feast. On such occasion, the clan's women and men would sing songs to bring honour and pride to the clan and their family. It is customary to offer meat and rice-beer to every person who partakes in his field's harvest. It may be noted that the latest such ceremony was performed by a retired teacher presently the 'Gaon bora' (chief of the village) Shree Imkongtemsu Jamir of Longsa village, Mokokchung, who achieved the honour of hosting mera nuja to the whole village of more than 600 households on 25th September 2020. In this event, the customs and practices were performed while hosting the whole village, including children, and partaking in the harvest along with a feast in the jhum field of about three hectares, which yielded more than 12 Mg of paddy along with many other crops.

4.2 Irregular or Emergency Ceremonies

Besides the regular ceremonies celebrated throughout the year, there are a number of irregular or emergency ceremonies that are associated with shifting cultivation observed by the Ao-Nagas. Whenever there are emergency circumstances due to natural calamities like floods, drought, insect or pest infestation of crops, such ceremonies are observed. They are mostly public ceremonies and they are as follows.

4.2.1 Mosumong

This is observed when withering or yellowing of the young paddy plants takes place due to drought or diseases. The priest declares a day to observe the among (ceremony) with prayer. On that day the farmers do not go to field but spend the day in the village observing anempong then the next day all the village men perform 'tzu-ok' (ceremonial community fishing) in the river.

4.2.2 Mesenphaba-Mong

This is a pest control ceremony. When the crops in the field are attacked by insects such as grasshoppers, stem-borers, tandi bugs etc. then the ceremony of mesenphaba-mong is performed by the farmers. ('mesen' means insects, 'phaba' means to catch, and 'mong' means ceremony). Here, a day is declared by the priest to observe the ceremony and on the appointed day all the male members go to the field and catch the insects that infest the crops. These insects are collected in a takzu (bamboo cup) which is then plugged with leaves and dropped into a river. If there are no big rivers it may be thrown outside the village fencing saying thus "*all the caterpillars and insects are captured and flow down to the 'molotsu' (ocean) never to come back to attack the paddy again*". All the people observe anempong on that day.

4.2.3 Tsungi-Tsungsang Mong

Sometimes the crops are destroyed by strong winds, hailstorms, floods, landslides etc. and this is a ceremony for protection from calamities. The field paths are also sometimes blocked by landslides or by fall of boulders or big trees, or sometimes weeding is made difficult due to heavy rain. In order to protect the lives of the farmers, the field and the crops from the dangers of all such natural calamities, a ceremony called tsungi-tsungsang mong is observed. On this day all the villagers are prevented from going to the field.

4.2.4 Menenmong

If any accident happens such as falling from a tree, burning to death during field burning, drowning in a river or stream, falling to death from a cliff etc., this leads to the dead body being disposed without allowing it to enter to the village and menenmong is observed by the family members and they perform kimoh to drive out the evil spirits.

4.2.5 Tsungi-Tsamong (Rain-calling ceremony)

Nagas, being predominantly agriculturalists, give great importance to rainfall. It is therefore natural to find many ceremonies designed to bring down rain in the form of welcome showers [10]. When there is no rain for a longer period and the crops are not growing well in the field, a day ceremony is observed in the village with prayer for good showers and healthy crop growth. Ceremonies for rain-making performed by the Ao-Nagas are (a) poisoning of the streams during ceremonial community fishing and (b) rain making through sacrifices done in the sacred forest.

Tzu-Ok (Ceremonial Community Fishing)

Only the men perform the customary community fishing by poisoning fish in the streams/rivers with plant extracts. A day is assigned during autumn after the paddy is harvested and the youngsters from the morungs collect fish poisoning plants the day before. These are species such as 'ruja' seeds of *Diospyros lanceifolia* (Ebenaceae) or roots of *Millettia* spp. (Fabaceae), or the bark and young fruits of *Juglans regia* (Juglandaceae) and take them to the stream before the next sunrise for traditional ceremonial fishing. After reaching the river bank, they take off the 'langtem' (loin cloth) worn by men and put the 'tsunglushi' (rain shields made of palm leaves *Caryota* spp., *Livistona jenkinsiana* (both Arecaceae) or *Pandanus* spp. (Pandanaceae)) over their heads pretending as if rain is falling. Then the old male priest is selected by the medicine man and pounds his bundle of the fish poison plants on an altar constructed in the side of the stream/river saying "*Is there any rain in the sky? Of course, there is, let it rain and never stop till the river is big enough to carry away an old man*". After this rite is over, all the men starts pounding their bundles of fish poison plants and allow their juices to flow into the stream/river. When the river gets sufficiently poisoned, the dead and dying fishes are caught in the ordinary way, with hands, fishing nets and bamboo baskets. Ovung et al. [11] described this community fishing by the nearby Lotha and Sumi tribes of Nagaland in more detail.

Rain Making Ceremony Through Sacrifices in the Sacred Forests

This ceremony is performed during the month of March or April and initiated by the arasentsur (medicine man) of the village by pulling out the apical shoot of a freshly cut plant of a Poaceae or Zingiberaceae plant from the sacred forest. The village priest performs the sacrifice ceremony to the sacred stones in the sacred forest and prays for rain on behalf of the whole village while all the villagers observe kimoh and no one is allowed to leave the vicinity of the village. Normally this is observed if the pre-monsoon is delayed, or when there is no rain for a long time. It is also predicted through the observation that when the priest or medicine man plucks the shoots of particular plants and shrubs and pronounces that the year will be a 'meken kum' (year of reduced rainfall) and then the rain-making ceremony will take place, and if the plucked shoot is healthy, juicy and fleshy it is predicted to that it will be a 'mezen kum' (year of normal rainfall) year. Prediction is also done by observing some particular occurrence of insect characters. For this ceremony, first of all, all the morungs of the village are repaired and cleared of their scrub, then five cocks are sacrificed with a prayer for rain-making near the 'kulemtong' (worship tree), this is usually a large tree either any one among *Alstonia scholaris* (Apocynaceae), *Erythrina arborescens, Erythrina stricta, Ficus benghalensis, Ficus elastica, Ficus racemosa* (Moraceae), *Quercus serrata* (Fagaceae), *Stereospermum chelonoides* (Bignoniaceae) or in the sacred forest near the village. The arasentsur examines the entrails of the cock and then predicts the ceremony.

5 Conclusions

Rituals may be prescribed by the traditions of a community, including a religious community. The Ao-Naga practices intense beliefs showing definite sign of being animistic in nature and polytheistic, worshiping several benevolent and malevolent spirits and gods at different times and occasions for different purposes. In the existence of supernatural powers in the universe, they perform animal sacrifices by offering pig, dog, fowl, eggs etc., to pacify their spirits and gods. There are a series of activities involved in ritual gestures like words, and objects, performed in appropriate places and performed according to a set order. The main characteristic features of rituals' are formalism. The Ao-Nagas believe in the existence of a supreme being, the creator of everything who is regarded as the fountain of all blessings for human beings and that he provides them protection and prosperity. By tradition, Ao-Nagas have always been insular, a democratic and sovereign entity. Therefore, it was inevitable that the religious beliefs touch all people in the villages in a connective way. The cycle of worship and sacrifices revolved around the practical activities of the people i.e. their agricultural basis and a firm attachment to culture, land and forest around them. The people depended on agriculture as their source of income, and also used forest resources for day-to -day requirements. Therefore, the need for a bountiful harvest from the fields and for protection from gods was of vital importance. Hence the religious ceremonies and rituals revolved around agriculture with a healthy respect for the environment. All the age-old forms of worship with sacrifice and chanting appropriate prayers continued through the oral tradition only. With the advent of Christianity, these particular type of ceremonies are no longer practiced, but the traditional celebrations are now performed with the Christian prayers during such events.

Acknowledgements Many thanks to the informants for this study Shri. Imkongniken (70 years), Mrs. Bendangkala (82 yrs) and Mr. Bendangyangba (86 years) of Khar village and Shri. Talimeri (85 years) and Shri. Tajennokcha (82 years) of Changki village of Mokokchung, Nagaland. Thanks also to Dr. Francis Brearley for helpful suggestions and edits to the chapter.

References

1. Ao T (1999) The Ao–Naga oral tradition. Heritage Publishing House, Dimapur, Nagaland, India
2. Mills JP (1926) The Ao Nagas. Macmillian and Co., London, UK
3. Smith WC (1925) The Ao Naga tribes of Assam: a study in ethnology and sociology. Macmillian and Co., London, UK
4. Dikshit KR, Dikshit JK (2014) Weather and climate of North-East India. In: Dikshit KR, Dikshit JK (eds) North-east India: land, people and economy. Springer, Dordrecht, The Netherlands pp 149–173. https://doi.org/10.1007/978-94-007-7055-3_6
5. Longkumer T (2022) Rethinking Ao Naga traditional religion. Transformation 39(2):99–110. https://doi.org/10.1177/02653788221099504
6. Longchar R (2016) Understanding the relation to between religion and environment of the Ao-Nagas through the ritual process. In: Dhanaraju V (ed) Regional environmental history: issues and concepts in the Indian subcontinent. Aakar Book, New Delhi, India, pp 194–208
7. Ramakrishnan PS (1992) Shifting agriculture and sustainable development: an interdisciplinary study from North-eastern India, UNESCO Man and the Biosphere Series, vol 10. UNESCO and Parthenon Publishing Group, Paris, France
8. NITI Aayog (2018) Report of working group III: shifting cultivation: towards a transformational approach. New Delhi, India
9. Nath AJ, Reang D, Sileshi GW (2022) The shifting cultivation juggernaut: an attribution problem. Glob Chall 6:2200051. https://doi.org/10.1002/gch2.202200051
10. Changkija S, Kumar Y (1996) Ethnobotanical folk practices and beliefs of the Ao-Nagas in Nagaland, India. Ethnobot Int J Soc Ethnobot 8(1&2):26–30
11. Ovung EY, Kithan L, Brearley FQ, Tripathi SK (2022) Indigenous community fishing practices in Nagaland, eastern Indian Himalayas. Sustainability 14(12):7049. https://doi.org/10.3390/su14127049

Innovative Shifting Cultivation and Other Agricultural Practices Conducted by the Indigenous Population of Mizoram, North-East India

Shri Kant Tripathi, Ramchhanliana Hauchhum, Etsoshan Y. Ovung, Ng. Somen Singh, David C. Vanlalfakawma, Keshav K. Upadhyay, Francis Q. Brearley, and Hnialum Lalraminghlova

Abstract In this chapter, we describe the shifting cultivation practices as practiced by the tribal communities in Mizoram, North-east India, a state characterised by steeply sloping land. As the sustainability of this practice is now being called into question, we outline new and innovative farming techniques used by Mizo farmers to diversify their produce, increase productivity, and conserve soil fertility. These include the use of bamboos and nitrogen-fixing species, or more complex inter-cropping approaches including agroforestry, integrated farming systems and plantations. We also describe some of the traditional and contemporary techniques for cultivating, processing and selling farm produce. Mizo society is undergoing rapid change and thus the synergies between traditional, improved and integrated farming systems may represent the most appropriate way forward.

Keywords Agroforestry systems · Bamboo-based farming · Changkham · Homegarden · Integrated farming systems · Jhum cultivation

---

S. K. Tripathi (✉) · R. Hauchhum · E. Y. Ovung · Ng. S. Singh · D. C. Vanlalfakawma · K. K. Upadhyay
Department of Forestry, Mizoram University, Aizawl, Mizoram 796004, India
e-mail: mzut041@mzu.edu.in

F. Q. Brearley
Department of Natural Sciences, Manchester Metropolitan University, Chester Street, Manchester M1 5GD, UK

H. Lalraminghlova
Department of Environmental Science, Mizoram University, Aizawl, Mizoram 796004, India

S. K. Tripathi and F. Q. Brearley (eds.), *Shifting Cultivation Systems*,
https://doi.org/10.1007/978-3-031-70388-1_3

1 Introduction

Mizoram is a rugged and hilly state located in North-east India sandwiched between Myanmar and Bangladesh; about half of the land area has slopes steeper than 20° [1]. It is known as the 'land of highlanders' and is among the eight 'sister states' found in North-east India, spread over an area of 21,081 km^2. The mean minimum and maximum temperatures range from 9 to 24 °C during winter and 24 to 32 °C during summer [2]. The region comes under the direct influence of the south-west monsoon, receiving an annual average rainfall of around 2500 mm [3]. The inhabitants of the state are known as the Mizos, a group that migrated from Myanmar since the seventh century [4]. The major Mizo tribes are: Hmar, Mara, Paihte, Pawi/Lai and other sub-tribes are Bru (Tuikuk) and Chakma with each tribe having their own culture and traditional customs.

The state of Mizoram was endowed with abundant natural resources which was the major food source for the Mizo population. In the past, they lived in close harmony with nature as a part of the natural ecosystem. Later, it is believed that they started the practice of shifting cultivation, the knowledge of which they inherited before they migrated to their current location in North-east India [2]. This agricultural system has been long and closely associated with the culture of Mizo society. In general, the Mizos believe in the conservation of nature, and biodiversity in particular, as they know that they depend on natural resources for their subsistence for food, medicine, fodder, fibre, etc. Therefore, over time, they developed various techniques to conserve biodiversity and soil fertility by ensuring a prolonged fallow period during the process of shifting cultivation to maintain the productivity of the land and enhance sustainability.

The practice of shifting cultivation and biodiversity conservation was balanced in earlier times because of low population densities, and the tribes of Mizoram remained undisturbed until 1826 during the British period [2]. However, in the last few decades, due to increasing population density, and despite efforts to conserve the tradition, agricultural productivity is declining due to shorter fallow periods that has created concerns regarding food security and environmental conservation [5]. This has been a major issue for the both the state and federal governments and government agencies have attempted to develop sustainable farming systems in the state. During the last three decades, various policies have been introduced by the government for this purpose, however, none of them have had major success due to various reasons [2]. Recently, the Government of Mizoram has introduced a New Land Use Policy as an alternative to traditional shifting cultivation which has had some impact in decreasing the area under shifting cultivation. With a change in government in 2018, this may lead to some further changes to the farming policies and consequences for farming practices. In spite of the major time investment in shifting agriculture, the output is still too low to fulfil the demand of the state's population. As such, import of food grain and other vegetables from neighbouring state remains high [6].

This chapter highlights the current scenario of shifting cultivation and its various operations, as well as new and innovative farming techniques used by Mizo farmers to increase productivity, conserve biodiversity and soil fertility, and increase the availability of resources with least environmental impact.

2 Shifting Cultivation in Mizoram: An Overview

Shifting cultivation is locally called 'jhum' meaning 'to group or work together'. This agricultural process is thought to have originated in the Neolithic period around 7000 B.C. [7] and is now widely practiced by millions of people inhabiting tropical moist forest areas of the world [8, 9]. It is the prevalent form of agriculture extending to South-east Asia and including North-east Indian states. Many Mizo historians claim that shifting cultivation was practiced long before settling in what is now known as Mizoram and it was the sole profession of the Mizo until the late nineteenth century [10, 11]. This practice still continues today due to the transfer of knowledge from generation to generation that provides an agricultural system with minimal initial costs. Furthermore, indigenous populations often lack adequate alternate livelihood options that can be both productive and economically viable [2]. Shifting cultivation has deeply impacted the socio-cultural aspects of Mizo society and many festivals are named and based on the various stages of this practice. Out of the 12 months of the lunar calendar, six months derive their name from the stages of shifting cultivation in the Mizo language.

Agriculture is the main occupation, and shifting cultivation continues to be the predominant practice, affecting as much as 6000 km^2, or about 28%, of the state land area. The Forest Survey of India estimated that around 0.38 million ha (c. 18 % of the state land area [12]) was influenced by shifting cultivation between 1987 and 1997. However, changes in the area under jhum fallow for the last 10 years have shown that there was a sharp increase in fallow area during 2008–09 to 2012–13 that may be related to changes in the government policy during that period [2].

Under jhum, farmers slash vegetation from approximately 0.5–2.5 ha of forest, depending upon availability, and burn it in situ after drying, followed by manual seeding and growing of rain-fed agricultural crops usually for one, or occasionally two, years depending on the length of fallow and soil fertility. The land is then abandoned to restore the soil fertility and farmers move to other locations for cultivation in subsequent years (Fig. 1) then returning to a given area on a rotational basis. Earlier, this agroecological system was productive and adequately balanced with a prolonged fallow period but, due to decreased fallow period, it is now becoming less sustainable [5]. During the past few decades, many policies related to shifting cultivation have been implemented in Mizoram, however, none of the practices promoted compete with traditional shifting cultivation in terms of productivity, economic feasibility and popularity; and thus these policies were not successful because of the lack of support among shifting cultivators (jhumias) [2].

Fig. 1 Shifting cultivation (jhum) cycle in Mizoram, North-east India

3 Farmers' Activities Under Shifting Cultivation

3.1 Allotment of Land for Cultivation by Village Elders/ Council

In the history of Mizo society, each village was living independently under the rule of the chief (lal). The chief of each village was the owner of the land with overall responsibility for the socio-economic welfare of the village inhabitants [2]. The chief undertook distribution of the land for cultivation and was assisted by jhum experts (ramhuals). The area of land allotted differed depending on the size of the family as well as the number of males available for agricultural work in each family. Widows were given a plot of land close to the village and bigger families were given a larger area of land. Currently, the village council is given the authority to arrange and distribute the land among the villagers by a lottery system. The land allotted should be from community land and the village council should be aware of the age of the fallow that will be distributed so that short fallow land is not included. The land allotted to jhumias gives no permanent rights to land tenure and it reverts back to the chief/village council after cultivation.

3.2 Slashing-and-Burning Activities

After the allotment of land, farmers prepare themselves by convening a small group of people to act together and conduct the various operations involved in the process of shifting cultivation. At the outset, slashing of the forest vegetation is carried out in December or January and the cut vegetation left on the ground to dry for one or two months. During slashing of the forest patch to be used for shifting cultivation, a fire line of about eight to ten metres width is maintained between the slashed vegetation and the adjoining forests thereby preventing the spread of fire to adjoining forest areas. The villages are generally encouraged to maintain leafy trees and bushes, whose branches can be used for quickly extinguishing fires at the fringes of jhum fields and adjoining forest areas, if needed. Prior to burning, the landowner, along with the village council members, generally undertake a 'field excursion' to inspect the fire lines, assess vulnerable points, and plan strategies for action in the event a fire escape. Penalties are generally imposed on individuals, or collectively on the entire community for deliberately or accidentally causing forest fires [13]. Burning activities usually start during March and continue through to April or occasionally May. In recent years, the Government of Mizoram notified that burning must be conducted before 15th March each year to avoid incidences of fire spread. Burning is always a busy period in a typical Mizo village with the entire village traditionally involved.

3.3 Cropping in Jhum Land

Shifting cultivation, or jhum, in Mizoram usually includes a number of seasonal as well as annual crops which are intercropped. Paddy, or rice, (*Oryza sativa*; Poaceae) is the principal crop cultivated, and other important crops include: cassava (*Manihot esculentum*; Euphorbiaceae), chillies (*Capsicum annuum*; Solanaceae), eggplant/brinjal (*Solanum melongena*; Solanaceae), okra (*Abelmoschus esculentus*; Malvaceae), maize (*Zea mays*; Poaceae), pineapple (*Ananas comosus*; Bromeliaceae), sesame (*Sesamum indicum*; Pedaliaceae), soyabeans (*Glycine max*; Leguminosae), squash (*Sechium edule*; Cucurbitaceae), sugarcane (*Saccharum officinarum*; Poaceae) and herbs such as mustard (*Brassica juncea*; Brassicaceae). In addition, ginger (*Zingiber officinale*; Zingiberaceae) and turmeric (*Curcuma longa*; Zingiberaceae) are frequently planted in recently burned sites as they grow well on steep slopes, and some horticultural crops are also grown. Different species and varieties of vegetables and horticultural crop occupy the same piece of land to meet the people's requirements; for example, mixed cropping of paddy with maize and ginger is commonly practiced. Sowing of paddy and other vegetables are mainly carried out by women after burning of forest patches or at the onset of the first monsoon shower generally between mid-April to the first week of May; the seeds are put in a small bag attached to their waist and sowing is carried out on the steep slopes. Numerous varieties of paddy are cultivated: short duration paddy varieties (e.g. bawngbuh, fazai, idaw, kungsei and thlanchhuah) are harvested in August/September, whereas long duration ones are harvested in October/November (e.g. biruchuk, buhpui, farel, maubuh, phulbuh, vuithup and zongam) [14].

3.4 Weeding Operation

Weeding is usually carried out three or four times during the cropping period depending on the growth and abundance of weeds; they are removed and mainly used for feeding cattle. The first weeding operation is done in May when weeds appear above the soil followed by a second weeding during June or July. Weeding is carried out until the paddy gains sufficient height, with a third weeding conducted during August; weeding usually stops by September although a fourth weeding may be carried out if necessary. The common weeds present are: *Ageratum conyzoides*, *Bidens pilosa*, *Cheilocostus speciosus Chromolaena odorata*, *Gynura nepalensis*, *Spilanthes acmella* (all Compositae), *Centella asiatica* (Apiaceae), *Knoxia sumatrensis* (Rubiaceae) and *Oxalis corniculata* (Oxalidaceae) among others. Weeding is usually done by women with the help of a hand-hoe (tuthlawh or chemkawm). Women also play a key role in the management of their plots by plant cultivation and caring for the plants to improve their growth and production [15].

3.5 Harvesting of Crops

Shifting cultivation involves the production of various kinds of crops and, therefore, harvesting occurs throughout the year (Fig. 2). During the harvesting season, women are responsible for collecting the vegetables upon their maturity and selling them in the local market to generate income. However, women are usually exempted from harvesting paddy as it is considered a hard job in which men play a more important role.

Fig. 2 Harvesting of shifting cultivation fields carried out by Mizo women

3.6 *Farmers' Strategies for the Next Cropping*

After the harvesting is over, farmers transport the crops back to their respective homes; it is tiring as they carry crops on their back over long distances uphill although, occasionally, they use vehicles if they are available. After the main harvest is an easier period and, whilst farmers wait for other crops to mature, they are free to spend time together at the fire places whilst discussing their harvest upon a cup of tea with local delicacies like taro (*Colocasia esculenta*; Araceae), tapioca (*Manihot esculenta*; Euphorbiaceae) and yam (*Dioscorea* sp.; Dioscoreaceae). During this period, they usually plan the farming strategies for the next year by considering the lessons learned from the previous cropping. During these discussions, farmers raise the burning issue of decreasing productivity due to the reduced fallow length to just three to five years in Mizoram [16]. Along with the reduction in crop production, increased soil erosional losses and decreased soil fertility are also key environmental consequences [5]. To overcome this issue and other concerns related to shifting cultivation, some farmers have adopted innovative farming practices with modifications to the traditional shifting cultivation techniques which are described below.

4 Innovative Practices in Shifting Cultivation and Related Farming Systems of Mizoram

4.1 *Changkham*

The gathering of pieces of burnt wood after the burning of a jhum field is called 'mangkhawh'. Laying the logs and bamboos along the slopes in rows every two to ten metres (depending upon the steepness of the slope) across the entire field, supported by the cut trunk bases or by pegs is known as 'changkham', which first emerged in 2012 in a few villages of Aizawl district. This is an effort by the Mizo farmers to improve the efficiency of the shifting cultivation system for sustained crop production by preventing soil erosion whilst allowing cropping in between the rows (Fig. 3) and conserving the fertile topsoil as well as retaining soil water more effectively.

Mizoram is a hilly state and under-developed region where farmers cannot afford the application of mechanical measures and have limited knowledge to implement new technologies to improve their farming system. Hence, the introduction of new farming techniques like changkham becomes an important step to sustain their agricultural production. Adoption of this approach may begin to transform shifting cultivation to settled agriculture and this practice has been enhanced by supplementing or replacing the wood and bamboo with hill-terraces or by planting nitrogen-fixing species like *Flemingia macrophylla*, *Tephrosia candida* or *Leucaena leucocephala* (all Leguminosae) [17].

Fig. 3 Changkham in jhum fields in Mizoram, North-east India

4.2 Bamboo in Jhum Cultivation

In addition to changkham practices, farmers of Mizoram have been trying to promote soil and water conservation through intercropping of agricultural crops like ginger or turmeric along the alleys formed by strips of bamboos (*Melocanna baccifera*; Poaceae) planted across the slopes of the jhum fields. This initiative allows the farmers to cultivate crops even under extremely steep slopes which are usually avoided due to their vulnerability to erosion. Usually, the farmer plants the bamboo on a small prepared jhum field in strips and allows it to grow along with other agricultural crops in between the strips (alleys) allowing two to three years of cropping. The underlying logic beneath this traditional initiative is that bamboos usually have a fast growth rate and can attain a substantial culm and root biomass within just a few years [18] which gives structural stability to the soil and thus allows cropping even under steeply sloping conditions.

4.3 Flemingia Macrophylla *in Jhum Cultivation*

Flemingia macrophylla is an important nitrogen-fixing indigenous shrub in Mizoram. This plant has also been successfully incorporated in jhum fields to serve several purposes, for example, conserving soil and water and improving soil fertility through enriching its nitrogen content. Farmers usually plant *Flemingia macrophylla* in a stripped pattern along with other agricultural crops in jhum fields. As well as adding valuable nitrogen to the soil, it also adds organic matter through litter deposition [19] and through decomposing shoot prunings [17]. In addition to litter deposition, it also gives structural stability to the soil thereby allowing prolonged periods of cultivation and increased output per unit area [20].

4.4 Agroforestry Systems

Various agroforestry systems have been established in the state. Plantations of teak (*Tectona grandis*; Lamiaceae) along with crops like rice and maize are the most commonly practiced of these [21]. Intercropping of soyabean with the bamboos *Melocanna baccifera* and *Dendrocalamus longispathus* (Poaceae) in degraded jhum land has been reported by Jha et al. [22]. The intercropping of ginger, turmeric and mustard with bamboos in degraded lands is noted to improve the soil characteristics [23]. Cultivation of paddy, maize and other vegetables with tung trees (*Vernicia fordii*; Euphorbiaceae) is another common agroforestry system adopted by the rural farmers. A recent study on plantations of fast-growing leguminous plants *Tephrosia candida* and *Flemingia macrophylla* along the contour lines of jhum fields found that they increased rice productivity and improved soil fertility due to an increase in soil organic matter content [24]. In addition, other leguminous plants such as pigeon pea *Cajanus cajan* (Leguminosae) and *Leucaena leucocephala* are also widely intercropped.

Coffee (*Coffea* sp.; Rubiaceae) based agroforestry systems are found in the Kolasib district of Mizoram established by the Department of Soil and Water Conservation, Government of Mizoram. Under this system, the existing trees are retained as shade trees, and coffee is allowed to grow in between these. The main tree species include *Bauhinia variegata* (Leguminosae), *Gmelina arborea* (Lamiaceae), *Erythrina variegata* (Leguminosae). *Magnolia champaca* (Magnoliaceae), *Schima wallichii* (Theaceae) and *Terminalia* sp. (Combretaceae). Similarly, a coffee-based agroforestry model is also practiced in Mamit district under the pilot project 'REDD+ Himalaya' [25]. In addition, a tea (*Camellia sinensis*; Theaceae) based agroforestry system is practiced on a large scale in Champhai and Serchhip districts of Mizoram in mildly-sloped areas.

Horticultural crops like banana (*Musa* sp.; Musaceae), pineapple, orange (*Citrus* × *sinensis*; Rutaceae) lemon (*Citrus* × *limon*; Rutaceae) etc. are also intercropped with vegetables as well as woody tree species. Pineapple and banana are the most common horticultural crops grown widely as they help in retaining soil moisture and also provide high economic returns. The Government of Mizoram has also encouraged farmers to focus on plantation species like *Anthurium* sp. (Aracaceae), broomgrass (*Thysanolaena latifolia*; Poaceae), dragonfruit (*Hylocereus undatus*; Cactaceae), rubber (*Hevea brasiliensis*; Euphorbiaceae) and livestock rearing. This has further resulted in a reduction in land area under shifting cultivation.

4.5 Homegardens

Homegardens are age-old practice in every Mizo village. There are two types of home garden systems in Mizoram viz. 'leipui' and 'chuktuah-huan'. Leipui are small patches of land, smaller than a jhum field, located in the vicinity of the village where

Fig. 4 Home gardens at Tanhril (Aizawl) Mizoram, North-east India

annual plants, mainly fast-growing vegetables are planted; they are analogous to a kitchen garden. Normally, senior citizens, who are still active, but not strong enough to work in the jhum field, focus on their leipui. Even though commonly practiced throughout the state, use of leipui has been poorly documented.

Another form of home garden system is the chuktuah-huan system, which is a contemporary term, originating in the late 1990s, meaning a garden conjoined with the home (Fig. 4). The primary objectives of these gardens are to produce item required for daily household consumption, and the surplus may be sold to the neighbouring families or in the market. These gardens comprise of woody trees, bamboos, horticultural crops and different vegetables which are managed by the owner family—mainly the female members thereof. In some chuktuah-huan, crops are mixed with orchards, forestry crops, pisciculture, floriculture, apiculture, livestock, piggeries or poultry rearing. Many authors regarded chuktuah-huan as the traditional form of homegardening in Mizoram [26, 27].

4.6 Bamboo-Based Farming Systems

Bamboo forests cover around 57% of the total area of Mizoram and most of the species are found between 400 and 1500 m altitude. The state possesses the greatest percentage of land area under bamboo forests compared to other Indian states. Bamboo forests are most abundant in the five northern districts although they are rarer in the eastern and south-eastern part of the state as these are at generally higher elevations. There are 35 bamboo species recorded from Mizoram and *Melocanna baccifera* covers about 95% of the total bamboo growing stock [28]; other common bamboo species include *Dendrocalamus longispathus*, *D. strictus*, *D. hamiltonii*, *Bambusa tulda*, *B. mizorameana* and *Schizostachyum dullooa* (all Poaceae). Bamboo has been clear-felled and burnt for jhum cultivation; however, it has also been recorded as an early colonising plant species in abandoned jhum land [29].

Fig. 5 Bamboo plantations on abandoned *jhum* land in Mizoram, North-east India

After the introduction of the state government's flagship New Land Use Policy, many beneficiaries opted for bamboo farming as an alternative to shifting cultivation through various practices such as the plantation of bamboo on lands following shifting cultivation (Fig. 5). The state government, in collaboration with Department of Forest, Environment and Climate Change, distribute and sell bamboo seedlings at subsidised rates for large scale plantation. This has encouraged the establishment of a bamboo-based industry producing, among others, laminated board, bamboo chips, compressed bamboo timber and bamboo charcoal that are exported both internally and internationally. This shows that bamboo-based farming possesses the scope for transforming land-use, acting as a source of livelihoods for rural populations, and becoming an alternative to shifting cultivation.

4.7 Integrated Farming Practices

The Government of Mizoram has formulated and implemented various policies to attempt to deal with land-use change induced by what they see as extensive unsustainable shifting cultivation. However, the identification and implementation of an ideal land-use system is still subject to much debate. In Mizoram, the undulating topography, extreme slopes and poorly stable soils/sediments, combined with the climatic conditions, lead to landslides and soil erosion. In spite of these unfavourable circumstances, the farmers and some landowners have established integrated farming systems involving the interactions among several components. For example, in the adjoining rural outskirts of the capital district (Aizawl), many of the land owners have established settled agricultural farms that integrate multiple components including agricultural crops (e.g. eggplant, cucumber (*Cucumis sativus*; Cucurbitaceae) and dragonfruit), medicinal plants (e.g. *Clerodendrum glandulosum*, *Elsholtzia* species, and *Mentha* species (all Lamiaceae)), valuable horticultural crops (e.g. *Areca catechu* (Arecaceae), passionfruit (*Passiflora edulis*; Passifloraceae) and avocado (*Persea americana*; Lauraceae) along with multipurpose trees such as *Parkia timoriana* (Leguminosae) and *Magnolia champaca* as well as livestock (poultry, pigs and cattle) and fuse them into single farming system. Deb [30] considers this a sustainable farming system where the components share a symbiosis among each other enhancing complex interactions, thereby optimising ecosystem functioning and developing a sustainable inter-dependent farming system while increasing the output per unit area. This type of farming system has only developed and evolved recently so realisation of its full potential is still developing—this may be partly attributed to the lack of sufficient private lands available, coupled with the requirement for capital investment during initiation. Nonetheless, this settled type of farming system can be practiced without shifting to another location since it involves a sustainable approach and it has been found to maintain the agro-ecosystem buffering capacity and prevent soil erosion as a result of the continuous vegetation cover throughout the year. In addition, mixed cropping and addition of organic manure (e.g. animal dung) increases soil fertility, especially of available nutrients like phosphorus, calcium and magnesium [31]. Therefore, this type of integrated settled farming system with the inclusion of economically valuable crops such as dragonfruit (Fig. 6) along with livestock as one of its main components may prove to be one of the most viable options for integrated farming systems in Mizoram. Recently, dragonfruit has been realised as an economically important plant by the farmers of Mizoram; it can fetch a price of Rs. 300–350 kg^{-1} and thus many of the land holders have devoted their lands to pure or integrated cultivation of this species (Fig. 6) with support from the state government.

The Department of Forestry, Mizoram University, under the leadership of Prof S. K. Tripathi, has undertaken proactive measures to establish integrated farming models under the Department of Science and Technology—Technology-based Socio-economic Development (DST-TSD) funded project entitled '*Establishing an indigenous innovative farming model for sustainable crop production and tribal empowerment in Mizoram*'. This initiative has resulted in the implementation of integrated farming models in two villages (Tachhip and Tanhril, both in Aizawl district). The comprehensive system comprises multiple components including livestock farming (involving the provision of cattle to farmers), polyhouse infrastructure, a hatchery, and a biofloc unit on both sites (Fig. 7). The livestock component ensures immediate economic gains to the farmers, while the complementary elements contribute

Fig. 6 Plantations of dragonfruit (*Hylocereus undatus*) in Mizoram, North-east India

Fig. 7 Polyhouse component of an integrated faming model in Mizoram, North-east India

to enhanced income generation and job opportunities for the farmers. An array of training programs and workshops were organised during the project and involved active participation from scientists, microbiologists, extension officers from fisheries, as well as bank managers for further enriching the scope of the initiative; over 200 farmers participated in these opportunities. The project also facilitated the distribution of high yielding varieties of seeds of different cash crops, e.g. amaranth (*Amaranthus* spp.; Amaranthaceae), bottle gourd (*Lagenaria siceraria*; Cucurbitaceae), chickpea (*Cicer arietinum*; Leguminosae), musk melon (*Cucumis melo*; Cucurbitaceae), okra, ridge gourd (*Luffa acutangula*; Cucurbitaceae), and fruit crops (like the apple (*Malus sylvestris*; Rosaceae) HRMN-99 variety for tropical areas) along with beehives. Additionally, biofertilisers were also provided to the framers for increasing their farm income. Notably, these multifaceted efforts reflect a commitment to scientific delivery to the farmers for sustainable agricultural practices well-suited to the landscape of the region with holistic community development and fostering of tribal empowerment.

5 Limitations to Shifting Cultivation in Hilly Regions

Indigenous communities around the world have been practising shifting cultivation for thousands of years as described above, however, still this practice still has limitations which include:

- Low productivity: Shifting agriculture is characterised by lower yields in comparison to other agro-ecosystems and, therefore, this system restricts farmers' income and food security options. On the other hand, the high diversity of crops in shifting cultivation fields supports agro-ecological resilience [32].
- Lack of access to markets: Most of the farmers are situated in rural areas with poor infrastructure leading to more difficult accessibility to markets, thus making it challenging for them to obtain an income from surplus produce or to procure materials needed for their farming activities [33].
- Land tenure: Shifting cultivators are allotted land for only a few years and therefore they are incentivised to extract the maximum from the land without any investment into it. Farmers do not have legal rights to hold the land beyond a certain limit and thus they do not plan for the future of that particular land holding [34].
- Loss of indigenous knowledge: Shifting cultivation was performed in the past by the involvement of the whole society which allowed the transfer of traditional knowledge and practices from one generation to the next. The practice is now becoming less common and often restricted to the older members of the community resulting in cultural erosion, loss of valuable cultural heritage and reduced wellbeing [35].
- Soil erosion and loss of soil fertility: Shifting cultivation can lead to soil erosion of up to 170 Mg ha^{-1} yr^{-1} [36], over time this leads to decrease soil fertility and crop productivity, although the erosion control techniques noted above can help to alleviate this problem. Similarly, cropping of tubers on steep slopes leads to loss of soil via harvesting [37] as well as opening of the soil and making it more vulnerable to further erosion, although this needs more research in the context of North-east India.

6 Various Agriculture Activities Derived from Traditional Knowledge

6.1 Collection of Healthy Seeds and Indigenous Systems of Storage for the Next Cropping

When harvesting crops, farmers select seeds from healthy and vigorous plants for the next cropping season. This selection for healthy seeds is based on the physical appearance such as shape and size. The traditional storage of seeds is simple and easy with seeds usually stored above the cooking stove (locally known as tap chhak) (Fig. 8). The seeds dry slowly and become free from viable fungi and other microbial spoiling agents. Rokhuma (in [15]) considers women to have 'historically had a deep scientific knowledge of the preservation of seed' as they are the traditional custodians of such knowledge.

Fig. 8 Storage of maize (*Zea mays*) seeds for next cropping season in a Mizo traditional kitchen

Fig. 9 Indigenous self-shopping market sheds established on roadsides by farmers selling their produce in Mizoram, North-east India

6.2 *Traditional Self-Shopping Sheds*

In Mizoram, farm produce are most commonly sold via community-established marketing sheds or to other private vendors. A traditional form of self-shopping sheds are now commonly being established on roadsides where the prices for the items on sale are advertised so that the travellers can make their purchase by simply choosing the desired item and depositing the correct amount in the box provided (Fig. 9), such is the sincerity of the market in Mizoram.

6.3 *Traditional Processing of Various Farm Produce*

In Mizoram, some of the farm products are now successfully processed in their place of origin and local patents are provided. For example, turmeric, a common spice, is cultivated extensively by the farmers of Reiek village, processed to turmeric powder

Fig. 10 Turmeric (*Curcuma longa*) plantations at Reiek village for turmeric powder production, Mizoram, North-east India

manually and sold in the markets. Turmeric is either grown along with a combination of other crops in jhum cultivation or monoculture plantations are established (Fig. 10). It is also often cultivated during the second-year cropping of jhum cultivation. In addition to turmeric, various value-added products varying from pickles to fruit drinks are processed, for example, amla (*Phyllanthus emblica*; Phyllanthaceae), mango (*Mangifera indica*; Anacardiaceae), banana inflorescence and bamboo shoots are processed into pickles, and lemon is locally processed to a ready-to-serve beverage. Post-harvest technologies and value addition have enhanced the products thereby improving the economic return by allowing longer periods of storage and increasing the availability of products during the off seasons.

6.4 Indigenous Threshing and Winnowing of Produce

Threshing and winnowing are complementary activities associated in the processing of rice grains after harvesting. Threshing refers to the loosening of the edible part of the grain by subjecting it to agitation and abrasion so as to remove the husks and is usually carried out manually by using a mortar and pestle-like structure framed out of a wood trunk with the pestle carved from poles of suitable tree species. After removal of the rice grains from the stalks, the Mizo farmers usually dry the rice grains in the sun to remove moisture, after which they are put in the mortar and pounded lightly with the pole to remove and loosen the edible part from the pericarp. The loosened rice grains along with the pericarp are then carried onto a platform raised above the ground, which is constructed using bamboos and poles, established especially for winnowing i.e. removal of the grains from the husks. In this traditional winnowing system, the farmers carry the processed grains onto the platform where they slide the basket gently allowing the rice to fall over the platform in an orderly fashion so that the wind can remove and separate the husks from the rice which then gets collected separately on the laid pher (woven bamboo mat).

6.5 Small to Medium Indigenous Water Storage Systems for Cropping

In the hills of Mizoram, ground water is hard to reach for extraction and therefore there is water scarcity during the drier months. Farming systems in Mizoram are usually rain-fed and farming becomes difficult under such conditions, so this has pushed the farmers to find alternatives to harvest water through various storage systems. Some of these indigenous water systems include small to medium fisheries, small tanks dug out and filled in with tarpaulin sheets or small raised tanks framed out of wood and bamboo, again made with tarpaulin sheets. These water storage structures have helped the farmers when the water availability to plants is reduced. Such structures are usually seen in settled agricultural farms and home gardens but are often also found established in the vicinity of many households in the hilly rural areas where water availability is poor.

7 Conclusions

When the Mizos migrated to their current locality in North-east India during the seventh century, they were bestowed with a lush green landscape enriched with a diversity of flora and fauna that was sufficient for leading their traditional lifestyle. However, they established shifting cultivation in the area, which led to profound changes in the landscape. Over the course of time, various changes have become apparent in the Mizo agricultural systems and thus their society as a whole, for example: (a) shortened jhum fallow period and poor forest regeneration, (b) decreased soil fertility and reduced agricultural productivity, (c) increase in private land holdings and social insecurity due to privatisation of land, (d) conversion of forest land into exotic plantations like teak, acacia (*Acacia auriculiformis*; Leguminosae), tung, *Jatropha curcas* (Euphorbiaceae), several bamboo species, red oil palm (*Elaeis guineensis*; Arecaceae), rubber, etc., (e) reduced land available for subsistence farming and cash crop cultivation, (f) loss of authority of traditional community institutions and (g) government encouragement for afforestation of jhum lands and consequently migration of the rural poor to urban areas.

Mizoram is still blessed with a huge diversity of bio-resources which are vital for the livelihoods of the rural population because of their high dependency on these resources. The conservation and efficient use of such resources are thus essential, and the principal responsibility of the broader community through biotechnological intervention and sustainable utilisation. The Government of Mizoram, supported by the central Indian government, has been trying to develop an efficient and ideal land-use system suitable for the hilly terrain in Mizoram but fully appropriate strategies are yet to be developed despite their efforts. Nonetheless, there have been many ingenious methods developed by the farmers in Mizoram which have contributed to the judicious use of forest lands, for example, the changkham technology which has been traditionally inherited and further evolved through the experience of the indigenous farming community. It is a method that is easy to implement and can preserve soil fertility, conserve water and promote more sustainable agricultural production. Such improvisation of traditional methods by building on the farmers' knowledge has proved to be an indication of efficient land management systems in the hills of Mizoram. In addition, many farmers use the technology adopted elsewhere to restore soil fertility and reduce soil losses due to erosion by establishing hedgerows of indigenous legumes and bamboos and other species for cropping on the steep slopes.

The growing population has catalysed rapid urbanisation and increased demand for food leading to increased rates of land-use change. In the Mizoram context, landslides and soil erosion are a common phenomenon in the majority of the steeply sloping areas which places pressures on the farmers for cultivation under such conditions. However, the identification and implementation of improved farming systems has, to some extent, prevented loss of land resources. The role of both traditional and integrated farming systems with appropriate management practices are thus important for ecoregions such as Mizoram which are challenged by lack of accessible agricultural lands and decreasing jhum fallow periods. This review was therefore inscripted with the motive to portray the traditional shifting cultivation practices augmented by the innovative farming practices and related activities performed by the farmers of Mizoram for enhancing agricultural production whilst promoting sustainability.

References

1. Monsang NP, Tripathi SK, Singh NS, Upadhyay KK (2021) Climate change and Mizoram: vulnerability status and future projections. In: Sati VP (ed) Mizoram: environment, development, and climate change. Today and Tomorrow's Printers and Publishers, New Delhi, India, pp 33–40
2. Tripathi SK, Vanlalfakawma DC, Lalnunmawia F (2017) Shifting cultivation on steep slopes of Mizoram, India: impact of policy reforms. In: Cairns M (ed) Shifting cultivation policies: balancing environmental and social sustainability. CABI, Wallingford, UK, pp 393–413. https://doi.org/10.1079/9781786391797.0393
3. Sati VP (2022) Mizoram: land and people. Today & Tomorrow's Printers and Publishers, New Delhi, India
4. Anon (2022) About Mizoram: history. https://mizoram.nic.in/about/history.htm. Accessed 17 April 2023

5. Grogan P, Lalnunmawia F, Tripathi SK (2012) Shifting cultivation in steeply sloped regions: a review of management options and research priorities for Mizoram state, Northeast India. Agrofor Syst 84:163–177. https://doi.org/10.1007/s10457-011-9469-1
6. Anonymous (2015) The study on development and management of land and water resources for sustainable agriculture in Mizoram in the Republic of India. https://irrigation.mizoram.gov.in/uploads/attachments/9e7a3373639b495387530345bce68b80/pages-114-id-jr15005-main-report-01.pdf. Accessed 11 Mar 2023
7. Das D (2006) Demystifying the myth of shifting cultivation: agronomy in the north-east. Econ Polit Wkly 41:4912–4917
8. Chao S (2012) Forest peoples: numbers across the world. Forest Peoples Programme, Moreton-in-Marsh, Gloucestershire, UK
9. Heinimann A, Mertz O, Frolking S, Christensen AE, Hurni K, Sedano F, Parsons Chini L, Sahajpal R, Hansen M, Hurtt G (2017) A global view of shifting cultivation: recent, current, and future extent. PLoS ONE 12:e0184479. https://doi.org/10.1371/journal.pone.0184479
10. Lalthangliana B (1998) Pi Pu Chhuahtlang: studies in Mizo culture and folktales. Hrangbana College, Aizawl, Mizoram, India
11. Lianthanga C (1999) Hmanlai Mizo nun. Mizoram Publication Board, Aizawl, Mizoram, India
12. Forest Survey of India (2000) State of the Forest Report. Forest Survey of India, Ministry of Environment & Forests, Government of India, Dehradun, India
13. Darlong VT (2002) Traditional community-based fire management among the Mizo shifting cultivators of Mizoram in northeast India. In: Moore P, Ganz D, Tan LC, Enters T, Durst PB (eds) Communities in flames: proceedings of an international conference on community involvement in fire management. Food and Agriculture Organization of the United Nations, Bangkok, Thailand, pp 119–124
14. Thansanga R (2000) Agriculture in Mizoram. Adland Publicity Pvt. Ltd., Culcutta, India
15. Hmingthanzuali PR (2009) Women's indigenous knowledge and relationship with forests in Mizoram. Asian Agri Hist 13:129–146
16. Tawnenga SU, Tripathi RS (1996) Evaluating second year cropping on jhum fallows in Mizoram, north-eastern India—phytomass dynamics and primary productivity. J Biosci 21:563–575. https://doi.org/10.1007/BF02703219
17. Fanai M, Gopichand B, Lalnunmawia F (2012) Intercropping of nitrogen-fixing tree species (NFTs) with agriculture crops as a sustainable farming system. Sci Vis 12:83–88
18. Tripathi SK, Singh KP, Singh PK (1999) Temporal changes in spatial pattern of fine root mass and nutrient concentrations in bamboo savannas in Indian dry tropics. Appl Veg Sci 2:229–238. https://doi.org/10.2307/1478986
19. Lalramliani CB, Lalnunzira C, Tripathi SK (2016) Litter decomposition and nutrient release pattern in two nitrogen fixing shrubs (*Flemingia macrophylla* and *Tephrosia candida*) growing in SALT farm in Lunglei district of Mizoram. Sci Vis 16:106–112
20. Zhu X, Liu W, Jiang XJ, Wang P, Li W (2018) Effects of land-use changes on runoff and sediment yield: implications for soil conservation and forest management in Xishuangbanna, Southwest China. Land Degrad Dev 29:2962–2974. https://doi.org/10.1002/ldr.3068
21. Lalramnghinglova JH, Jha LK (1996) Prominent agroforestry systems and important multipurpose trees in farming system of Mizoram. Indian For 122:604–609. https://doi.org/10.36808/if/1996/v122i7/6395
22. Jha LK, Lalnuntluanga, Marak C (2004) Study on the growth performance of bamboo species *Melocanna baccifera* and *Dendrocalamus longispathus* along with crop (*Glycine max*) in degraded jhum land of Mizoram. Indian For 130:1071–1077. https://doi.org/10.36808/if/2004/v130i9/2122
23. Tewari S, Banik RL, Kaushal R, Bhardwaj DR, Chaturvedi O, Gupta A (2015) Bamboo based agroforestry systems. In: Kaushik S, Singh YP, Kumar D, Thapliyal M, Barthwal S (eds) Bamboos in India. ENVIS Centre on Forestry, Forest Research Institute, Dehradun, Uttarakhand, India, pp 261–284
24. Lalramliani CB (2017) Comparative study of impact of shifting cultivation and settled agriculture on soil fertility in Aizawl and Lunglei districts of Mizoram. PhD Dissertation, Mizoram University, India

25. Rawat RS, Arora G, Rawat VRS (2019) Implementation of REDD+ activities in North-Eastern states of India: a case study from the state of Mizoram. Indian For 145:871–878. https://doi.org/10.36808/if/2019/v145i9/148720
26. Sahoo U, Jeeceelee L, Lallinrawna S, Muthukumaran RB (2015) Effect of *Citrus reticulata* Blanco leaf extract on seed germination and initial seedling growth parameters of five home garden food crops. J Exp Biol Agric Sci 3:517–528. https://doi.org/10.18006/2015.3(6).517.528
27. Barbhuiya AR, Sahoo UK, Upadhyaya K (2016) Plant diversity in the indigenous home gardens in the Eastern Himalayan region of Mizoram, Northeast India. Econ Bot 70:115–131. https://doi.org/10.1007/s12231-016-9349-8
28. Anonymous (2010) Bamboos of Mizoram. Environment and Forest Department, Government of Mizoram, Aizawl, Mizoram, India
29. Hauchhum R, Tripathi SK (2017) Rhizosphere effects of *Melocanna baccifera* on soil microbial properties under different fallow phases following shifting cultivation. Int J Plant Soil Sci 17:34493. https://doi.org/10.9734/IJPSS/2017/34493
30. Deb S (2020) Traditional agroforestry systems of Northeast India. In: Roy N, Roychoudhury S, Nautiyal S, Agarwal SK, Baksi S (eds) Socio-economic and eco-biological dimensions in resource use and conservation: strategies for sustainability. Springer Nature, Switzerland, pp 103–115. https://doi.org/10.1007/978-3-030-32463-6_5
31. Ovung EY (2016) Impact of different land use systems on soil physico-chemical properties. MSc Dissertation. Mizoram University, Aizawl, India
32. Isbell F, Adler PR, Eisenhauer N, Fornara D, Kimmel K, Kremen C, Letourneau DK, Liebman M, Polley HW, Quijas S, Scherer-Lorenzen M (2017) Benefits of increasing plant diversity in sustainable agroecosystems. J Ecol 105:871–879. https://doi.org/10.1111/1365-2745.12789
33. Paul S, Das TK, Pharung R, Ray S, Mridha N, Kalita N, Ralte V, Borthakur S, Burman RR, Tripathi AK, Singh AK (2020) Development of an indicator based composite measure to assess livelihood sustainability of shifting cultivation dependent ethnic minorities in the disadvantageous Northeastern region of India. Ecol Indic 110:105934. https://doi.org/10.1016/j.ecolind.2019.105934
34. Thanga JLT (2019) Institutional changes in jhum land tenure system in Mizoram. Mizoram Econ Rev 3(2):20–38
35. Pandey DK, Dubey SK, Verma AK, Wangchu L, Dixit S, Devi CV, Sawargaonkar G (2023) Indigenous peoples' psychological wellbeing amid transitions in shifting cultivation landscape: evidence from the Indian Himalayas. Sustainability 15:6791. https://doi.org/10.3390/su15086791
36. Saha R, Mishra VK, Khan K (2011) Soil erodibility characteristics under modified land-use systems as against shifting cultivation in hilly ecosystems of Meghalaya, India. J Sustain For 30:301–312. https://doi.org/10.1080/10549811.2011.531992
37. Kuhwald M, Busche F, Saggau P, Duttmann R (2022) Is soil loss due to crop harvesting the most disregarded soil erosion process? A review of harvest erosion. Soil Tillage Res 215:105213. https://doi.org/10.1016/j.still.2021.105213

Innovative Indigenous Technologies for Sustainable Mountain Farming Systems in North-East India

Sapu Changkija, H. C. Hombegowda, Chowlani Manpoong, Sanjay Kumar Ray, and Francis Q. Brearley

Abstract North-east India is biologically and culturally diverse. In this region, shifting cultivation, known locally as jhum, is the dominant agricultural practice in which a diverse suite of plant crop species are grown. Additionally, in many locations, unique traditional cultivation practices based on Indigenous technologies have evolved. In this chapter, we describe some of these practices, namely: alder-based shifting cultivation, bun cultivation, zabo and Apatani fishery-cum-rice cultivation, agroforestry, and homegardens. We show how these Indigenous farming systems are ingeniously designed and sustainable as they are based around forest, soil and water conservation and the judicious recycling and reuse of resources. We explore the role of these practices in contributing to the UN Sustainable Development Goals under future climate change.

Keywords Agrobiodiversity · Biochar · Fisheries · Indigenous innovations · Integrated farming systesm

---

S. Changkija
Department of Genetics and Plant Breeding, Nagaland University, Medziphema, Nagaland 797106, India
e-mail: sapuchangkija@gmail.com

H. C. Hombegowda
Indian Council of Agricultural Research (ICAR)—Indian Institute of Soil and Water Conservation, Research Centre, Sunabeda, Odisha 763002, India

C. Manpoong
Department of Crop Production, Faculty of Agricultural Sciences, Arunachal University of Studies, Namsai, Arunachal Pradesh 792103, India

Department of Forestry, Wildlife and Environmental Sciences, Guru Ghasidas Vishwavidyalaya, Bilaspur, Chhattisgarh 495009, India

S. K. Ray
Indian Council of Agricultural Research (ICAR) Research Complex for NEH Region, Tripura Centre, Lembucherra, West Tripura, Tripura 799210, India

F. Q. Brearley (✉)
Department of Natural Sciences, Manchester Metropolitan University, Chester Street, Manchester M1 5GD, UK
e-mail: f.q.brearley@mmu.ac.uk

S. K. Tripathi and F. Q. Brearley (eds.), *Shifting Cultivation Systems*,
https://doi.org/10.1007/978-3-031-70388-1_4

1 Introduction

During human history, many of the communities living in tropical regions with acidic infertile soils have developed shifting cultivation to make best use of their ecosystems for agricultural production [1]. The basic premise is that a small area of land is cleared and burned (Fig. 1), and subsequent subsistence crop production occurs for one or two years before the land is allowed to fallow, during which secondary forest develops over a number of years before being re-cleared [2–7]. Burning of vegetation biomass increases the availability of essential nutrients and reduces soil acidity paving the way for crop production. In India, this practice is dominant in the North-east region as well as some pockets in the Eastern Ghats. Agriculture serves as the main source of livelihood for the majority of the populations in these regions and is linked with their socio-economic and cultural life and closely connected to the rituals and festivals that take place during various key stages of cultivation. One of the most remarkable features of shifting cultivation (known as jhum in the North-east) is that a large number of cereals, vegetables and fruits can be grown in one jhum field; it is common to find a high level of diversity of 30 or more non-rice (*Oryza sativa*; Poaceae) staple crops maintained in a single field of usually less than one hectare [8, 9]. There is also a remarkably high genetic diversity of varieties and landraces of the main crop species [10–12]. This mixed cropping and multi-layered structure are basic features of shifting cultivation making it into a 'biodiversity hot-spot' in the village community landscape (Fig. 2). However, to overcome the constraints to low crop production, shifting cultivation has undergone many localised modifications to soil and crop management practices for improving soil fertility and maintaining productivity.

The Indigenous traditional shifting cultivation is a complex farming system. The dynamics of shifting cultivation have led to the belief that as long as population density is low, rotational shifting cultivation is a sustainable system. This is because if the fallow cycles are maintained at an appropriately long period in order to regenerate

Fig. 1 Burning of vegetation for shifting cultivation field preparation in Arunachal Pradesh, North-east India

Fig. 2 Shifting cultivation landscape in Nagaland, North-east India. Photo courtesy of Dr Etsoshan Ovung

the vegetation biomass and nutrient capital, it does not result in degradation of the ecosystem. Nevertheless, soil erosion can be problematic, particularly in the early stages of cultivation during the onset of the monsoon and prior to the crop having covered the soil sufficiently to ameliorate its erosive effects [13]. To this end, soil conservation strategies are practiced such as bamboo splits placed randomly across the slope of a shifting cultivation field: this can reduce soil loss by more than 80% compared with a typically-managed field (50.7–70.1 Mg ha^{-1} yr^{-1}) in experimental studies [14]. Similarly, echo (Fig. 3) is a traditional soil conservation practice used by farmers in Wokha district and other parts of Nagaland. This entails placing logs or other locally available items across the slope at random that last for up to three years. Again, compared with a typically-managed shifting cultivation field, experimentally placing logs across the slope reduced soil loss by 60–65% [14]. Terracing as a form of cultivation to reduce soil erosion is not as common as shifting cultivation, although there are some terraces of both dry and wet rice found in Nagaland [15].

This chapter has been written with the objective of overviewing traditional practices of agricultural land-use in North-east India with a more detailed focus on some of the agricultural innovations developed in the region that are often complex in

Fig. 3 Echo traditional soil conservation techniques in Nagaland, North-east India. Photo courtesy of Dr Etsoshan Ovung

nature and, in some cases, involve cropping of plants along with rearing of animals in a mixed agricultural setting.

2 Alder-Based Shifting Cultivation

Alder-based shifting cultivation is a farming system practiced by a number of tribes in Nagaland, particularly the Angami, Konyak and Sema. The technique is most well-known from the village of Khonoma in Kohima district [16, 17], but is also practiced in other locations in Nagaland [18]. The alder tree, *Alnus nepalensis* (Betulaceae), locally called 'rupro' by the Angami, 'entong' by the Ao and 'young' by the Konyak, is a nitrogen-fixing species and thus contributes to maintaining soil fertility. Cairns et al. [16] provide an interesting and valuable history of the cultivation of the species and hypothesise that the historical warring between Naga tribes led to security concerns and thus intensification of land-use was promoted as farmers were not able to spend time in less well-protected field locations. Jhum fields planted with alder are usually managed in four-year cycles, with two years of cropping among the alder trees followed by two or more years of fallow to recuperate the soil and allow the canopies to expand [16, 19]. In this system, the alder trees are pollarded at a height of about two meters from the ground to obtain a head of new shoots, which are nurtured during the jhum period. Farmers pollard the trees on a rotational basis with the leaves and small twigs added to the soil after being burnt; the branches cut are used or sold as fuel wood. Rathore et al. [20] show how the total litter nitrogen yield is proportional to the number of alder plants, increasing up to 185 kg N ha^{-1} at 625 trees ha^{-1}. Farmers cultivate the fields with various crops and vegetables in this system, the main intercropping components are Job's tears (*Coix lacryma-jobi*; Poaceae), maize (*Zea mays*; Poaceae) and rice, and the sequential crops are cowpea (*Vigna unguiculata*; Fabaceae), millet (*Eleusine coracana*; Poaceae), perilla (*Perilla frutescens*; Lamiacaee), potato (*Solanum tuberosum*; Solanaceae), soybeans (*Glycine max*; Fabaceae), taro (*Colocasia esculenta*; Araceae), and various other vegetables growing among the alder trees [21]. Additionally, new alder saplings are also planted and left to mature during the fallow period to be pollarded during the next cultivation cycle.

Alder increases soil fertility by fixing atmospheric nitrogen through nodules containing *Frankia* bacterial species [22, 23]. Giri [24] additionally reported free-living nitrogen-fixing bacteria from alder jhum fields. Furthermore, there are suggestions that alder trees can contribute to rock weathering, thus making a source of recalcitrant nutrients more available in the ecosystems [25]. Mishra et al. [26] showed that soil carbon and nitrogen were greater under alder-based jhum compared to more typical jhum cultivation systems and that these differences were still seen to 60 cm soil depth, particularly for nitrogen which was markedly greater under alder-based shifting cultivation; there were no differences in soil phosphorus or potassium. Similarly, Kalidas-Singh et al. [19] showed how soils under alder-based shifting cultivation were more fertile than those under more 'typical' shifting cultivation practices across a broad range of fertility measures.

3 Zabo Farming System: Integrated Farming System with Fisheries, Livestock and Crops

The zabo farming system is practiced mostly by the Chakhesang tribe in Nagaland and Mao tribe in Manipur, it is also known as dzüdü or ruza. This farming system has a combination of forest and agriculture, mainly rice, incorporating livestock and fisheries. It is thus an integrated farming system with sustainable utilisation of rainwater and maintenance of soil fertility using organic sources of crop nutrition. The zabo system is located on a slope having forested land at the top of the hill, a silt retention tank followed by a water harvesting tank-cum-fishery below the forest and livestock yards (cattle, buffalo, goat/duck/poultry) and paddy-cum-short duration fishery at the foot of the slope with common carp (*Cyprinus carpio*; Cyprinidae) and Asian snakehead (*Channa striata*; Channidae) as the main fish species [27–29]. Some farmers also cultivate snails as an additional income source [29]. In some cases, when a sufficient water source is not available, the run-off from the catchment area is drained directly to the paddy fields. The water from the harvesting tank is diverted to irrigate the paddy field whilst passing through the livestock yards to collect nutrient-rich animal excreta used to fertilise the rice. The water loss and percolation from the hill slope terraces is controlled through soil compaction and the use of various materials like paddy straw, stones, etc. In this system, crops are also cultivated on the bunds and therefore soil erosion is minimised. Farmers may often maintain a small vegetable garden where they do not use any chemical fertilisers or pesticides, thus growing their vegetables organically. They manage the soil fertility by judicious crop rotation and incorporation of farmyard manure. Rice yield is 1.4–2.5 Mg ha^{-1} yr^{-1} [27, 28] with 40–80 kg fish ha^{-1} yr^{-1} [28]. The cropping pattern is paddy-cum-short duration fishery and snail rearing in summer and, in winter, various vegetables, such as cabbage (*Brassica oleracea*; Brassicaceae), mustard (*Sinapis alba*; Brassicaceae), potatoes, radish (*Raphanus sativus*; Brassicaceae), tomatoes (*Solanum lycopersicum*; Solanaceae) etc. are cultivated. The zabo farming system has inspired model integrated farming systems that incorporate aspects of water management. The zabo farming water harvesting and management system has served as an inspiration for model integrated farming systems [30–32] because it employs a technique for water harvesting and recycling as well as a conservation method to manage soil fertility and control soil erosion combined with the cultivation of various crops in tandem with animal rearing and, often, fisheries.

4 Bun Cultivation: Raised Beds Incorporating Biochar

Bun cultivation is one of the Indigenous agricultural practices of the communities in Meghalaya. Bun is a method of cultivation of tuber crops and vegetables in a series of raised beds where dried plant biomass, including weeds, obtained during slashing of the vegetation as well as lower branches from trees in the surrounding forest are piled in raised beds (around 1 m width by 4–7 m in length) and topped with a layer of soil of about 10 cm depth followed by burning of the covered vegetation before planting of crops (Fig. 4). The biomass is burnt under a limited oxygen environment to convert it into charcoal (biochar) through pyrolysis—this prevents the volatilisation of many nutrients that are, instead, returned to the soil. Around 6–10 Mg of biomass are burnt per hectare leading to up to 3 Mg of biochar being produced within the same area [33]. After burning has finished (up to a few days), crops are planted in the raised beds. Enrichment of the fields with biochar has a positive effect on soil physico-chemical properties. Amelioration of the acidic soil pH is the most significant change brought about by the biochar application which is attributed to its alkaline nature and the release of base cations that have a high buffering capacity. This method also facilitates nutrient retention and inhibits nutrient leaching especially nitrate [34] due to microaggregates that increase the porosity and surface area of added biochar. Through controlled burning at a moderate temperature, more nutrients are returned to the soil instead of being lost through volatilisation as would be the case with an open burning method with high temperatures. Furthermore, the raised beds allow free draining of excess water between them. Hombegowda et al. [33] showed increases in key soil nutrients by 50–100% in most cases after burning. Farmers also perceive a reduction in insect pests and weeds under the bun cultivation system [35]. The land is cultivated for up to three years and then remains fallow for a similar period of time [36]. Hombegowda et al. [33] further showed an increase in crop yields of ginger (*Zingiber officinale*; Zingiberaceae) and potato of about 30% and rice of about 70% under bun cultivation compared to shifting cultivation. This is a good increase in crop productivity when compared with the mean reported in the meta-analysis of Jeffrey et al. [37] of 10% with biochar application across a broad range of agricultural systems. Prokop et al. [36] found that bun cultivation improved soil quality under flat topographical conditions through the regular supply of nutrients from organic matter. However, this pattern was not seen when the cultivated land was steeply sloped and subjected to overland flow and soil erosion. This bun cultivation approach uses a form of biochar which is considered as having numerous benefits to agricultural production as well as sequestering carbon in soil over the long-term. However, the benefits are not consistently seen in all cases, as noted above, so the individual situation should be considered in each case [38]. Current innovations to this technique are the use of turf as a source of biomass rather than tree branches which may be becoming more scarce [39] and recent cultivation of cash crops using this technique [40].

Fig. 4 Bun raised beds covered with earth in preparation for burning and biochar formation in Meghalaya, North-east India

5 Apatani Rice and Fish Co-cultivation

The Ziro valley in Arunachal Pradesh, mainly occupied by the Apatani tribe, is well known for its paddy-cum fish-cultivation practice, known locally as ajii-nguyi, that covers about half of the land area [41–48]. Cultivation of rice and fish in combination has existed in parts of South-east Asia for 1200 years [49]: the presence of fish in the flooded rice fields helps control weeds and pests as well as providing nutrients from their excreta. The paddy-cum-fish cultivation in the Ziro valley is the primary occupation which provides the basic livelihood of the villagers and is considered an economically valuable farming practice as sufficient rice is produced to be exported. Importantly, the rice-fish culture increases rice yield and incomes by up to 30% [45]. In this system, paddy fields are irrigated by diverting streams flowing downhill such that fields are submerged to 10–20 cm and connected through channels to maintain the appropriate water depth. The fish, mostly small carp (Cyprinidae) species, and particularly common carp *Cyprinus carpio* [46], feed on plankton within the paddy fields; deeper trenches (c. 40 cm) are dug in parts of the paddy field in which fish can shelter. The first batch of fish fry is usually stocked before the transplantation of the rice seedlings. Two crops of rice are grown (early- and late-yielding) with the late-yielding variety grown closer to the village where nutrient status is greater [41] as household organic wastes are added to the fields. Additionally, millet (*Eleusine coracana*; Poaceae) is grown on the bunds between paddy fields and used to make 'o' which is a culturally important drink [50]. After harvesting, rice stubble is left in the field and augmented with other organic wastes from the villages such as animal dung, ash and kitchen refuse [43]. Additionally, *Azolla* (Salviniaceae) and *Lemna* (Araceae) species are allowed to grow in the submerged fields as nitrogen fixers [42]. Kanwat et al. [51] outline the traditional approach to controlling the rice insect pest *Leptocorisa acuta* (Alydidae) by attracting them with malodorous dead fish and animals. Production is around 200–500 kg fish ha^{-1} yr^{-1} and 3–5 Mg rice ha^{-1} yr^{-1} [41, 42, 47] and millet yield is around 100 kg ha^{-1} yr^{-1} [41]. A number of studies have examined the economics of this production system [44, 47] and this economic approach could be fruitfully extended to other farming systems noted in this chapter. In recognition of this unique agriculture practice, the Apatani valley is a cultural landscape included on India's tentative list for World Heritage status [52].

6 Agroforestry and Homegardens

In agroforestry systems, socially, culturally and economically valued species are grown in mixtures. The key characteristic feature of agroforestry systems is the growth of complementary species that maximise resource usage [53, 54]. The maintenance of tree-based vegetation cover helps to prevent soil erosion and the high leaf litter return maintains soil fertility. Agroforestry systems can be fruit-tree based, bamboo-based or timber-tree based [55] and are often a combination of these; the systems are usually highly diverse with numerous species grown in various combinations (Table 1).

Table 1 Examples of diverse agroforestry practices in North-east India

Key species	Location	Reference
Agarwood (*Aquilaria malaccensis*) homegardens	Assam	Saika and Khan [56]
Alder (*Alnus nepalensis*; Betulaceae) and cardamom (*Amomum subulatum*; Zingiberaceae) agroforestry	Sikkim	Sharma et al. [57]
Arecanut (*Areca catechu*; Arecaceae) and banana (*Musa* sp.; Musaceae) agroforestry	Tripura	Mathur and Bhattacharya [58]
Betel (*Piper betle*; Piperaceae)-based agroforestry	Barak valley (Assam)	Reang et al. [59]
Himalayan pine (*Pinus wallichiana*; Pinaceae) and bamboo (*Phyllostachys bambusoides*: Poaceae) agroforestry	Ziro valley (Arunachal Pradesh)	Tangjang and Nair [60]
Oil palm (*Elaeis guineensis*; Arecaceae)-based agroforestry	Mizoram	Ahirwal et al. [61]
Pineapple (*Ananas comosus*; Bromeliaceae)-based agroforestry	Barak valley	Reang et al. [59]
Toko palm (*Livistona jenkinsiana*; Arecaceae) agroforestry	Arunachal Pradesh	Sarangi et al. [62]

Homegardens with complex vegetation patterns and a notably high diversity of local plant species act as an agroforestry system found immediately around homesteads, which diversify the livelihood options for the Indigenous people [63]. They are used to supplement jhum crops with a broad range of fruits, vegetables, spices, medicinal plants, multi-purpose trees, and sometimes cash crops [64]. The complex structure and highly diverse species composition depends both on their position in the overall farming system and on livelihood strategies; the size of a typical home garden is less than 0.25 ha. Species diversity in homegardens is notably high [65] with a broad range of multi-functional species. Boundary hedges may be formed by, for example, *Bauhinia* (Fabaceae) species as a multi-purpose tree species that provides nitrogen fixation, fencing, and beautification. Finally, homegardens have important cultural values and provide social functions through the provision of gifts in the form of fruits, leaves or products for religious or medicinal purposes.

7 Discussion

Agroecosystems in the North-east region of India are indigenously designed and modified by the rural farmers to meet their livelihood needs and, of key importance, are supported by the forest ecosystems. In this chapter we have considered the traditional ecological knowledge of the tribal communities as contributing towards sustainability in this hilly, and therefore ecologically vulnerable, Himalayan region. Due to the hilly terrain and rugged topography, the scope for use of mechanised techniques in this region is limited leading to mostly small-scale farming practices, many of which are for subsistence use only. Small land holdings and a lack of credit opportunities have restricted the choices of the farmers in terms of increasing their yield, yet shifting cultivation remains a practice with immense cultural value. There are also a number of agricultural systems which, with appropriate innovations, management and encouragement, may have tremendous potential for sustainable improvement without disturbing the traditional farming systems and the soil or forest ecological processes [21]. We now consider what lessons can be learnt from these systems in order to support food security and agricultural sustainability in North-east India and how they might contribute to climate change resilience in the near future.

Many of the agricultural practices described in this chapter are 'organic by default' and based on the use of resources solely available in the immediate area; furthermore, they are highly cyclical systems with judicious recycling of nutrients supporting a more 'closed' agro-ecosystem. Additionally, ecological synergies among components in the agro-ecosystems are harnessed, such as that between rice and fish when they are co-cultured to increase the yield of both of them [49]. Such approaches minimise the inputs of external nutrients such as inorganic fertilisers which can be expensive for shifting cultivation farmers as well as potentially ecologically harmful [66]. There are synergies between shifting cultivation and organic agriculture [67] and suggestions to certify shifting cultivation produce as organic have been raised [68] with this approach capitalising on an increasing demand for organic foods in

India [69]. Furthermore, many of these agroecosystems could be considered Globally Important Agricultural Heritage Systems [70] and, indeed, have been suggested for World Heritage status, in some cases [52]. These proposed global designations all give credence to the uniquely valuable nature of some of these agricultural systems.

The climate is changing in North-east India with increased water stress and more erratic rainfall [71, 72] which is important because water plays a key role in these rain-fed agro-ecosystems. However, there are a number of ways in which the traditional agricultural practices conducted in North-east India present resilience against the currently increasing effects of climatic change. As examples, the raised beds used under bun cultivation can mediate the effects of excess water by draining excess runoff, and the rainwater harvesting techniques used in zabo farming can mitigate against low rainfall. Furthermore, the additions of organic matter in the alder-based shifting cultivation and bun cultivation systems promote soil water retention whilst also providing additional organic forms of nutrients to the crops. Agroforestry systems can also promote soil organic matter accumulation through natural litter decomposition, as can the use of other intercrops and green manures [73]. Rice-fish cultivation is more efficient in the production of rice whilst reducing nitrogen losses to the environment (although likely increases methane emissions) [74]. It is known that diverse agricultural mixtures are more resistant to climate shocks, whilst maintaining high yields [75] and thus shifting cultivation systems have a role in supporting agro-ecosystem resilience. In addition to developing climate change resilience, these agro-ecosystems can help mitigate climate change through a number of mechanisms. The low external fertiliser inputs lead to less nitrous oxide (a potent greenhouse gas) emissions [76] and the growth of trees in tree-based agroforestry systems and jhum fallows (i.e. secondary forests) can sequester carbon dioxide over longer time frames [77]. Finally, the use of biochar in bun cultivation can also contribute to soil carbon sequestration [78].

The understanding of the functioning of agro-ecosystems gained from these traditional practices can provide valuable knowledge to support a number of the Sustainable Development Goals, in particularly 2 (Zero Hunger) and 15 (Life on Land) with additional contributions to 1 (No Poverty), 6 (Clean Water and Sanitation) and 13 (Climate Action). In conclusion, and as noted by Antonelli [79], due to the diverse, resilient and circular nature of many agricultural systems in North-east India, we need to strongly bear in mind that 'Indigenous knowledge is key to sustainable food systems' and harness and promote this knowledge for more sustainable future endeavors.

Acknowledgements We thank Dr Etsoshan Ovung for providing photographs for this chapter.

References

1. Heinimann A, Mertz O, Frolking S, Christensen AE, Hurni K, Sedano F, Parsons Chini L, Sahajpal R, Hansen M, Hurtt G (2017) A global view of shifting cultivation: recent, current, and future extent. PLoS ONE 12:e0184479. https://doi.org/10.1371/journal.pone.0184479
2. Das D (2006) Demystifying the myth of shifting cultivation: agronomy in the North-east. Econ Polit Wkly 41:4912–4917
3. Grogan P, Lalnunmawia F, Tripathi SK (2012) Shifting cultivation in steeply sloped regions: a review of management options and research priorities for Mizoram state, Northeast India. Agrofor Syst 84:163–177. https://doi.org/10.1007/s10457-011-9469-1
4. Kerkhoff E, Sharma E (2006) Debating shifting cultivation in the Eastern Himalayas: farmers' innovations as lessons for policy. International Center for Integrated Mountain Development, Kathmandu, Nepal
5. NITI Aayog (2018) Report of working group III: shifting cultivation: towards a transformational approach. New Delhi, India
6. Ramakrishnan PS (1992) Shifting agriculture and sustainable development: an interdisciplinary study from North-eastern India, UNESCO Man and the Biosphere Series, vol 10. UNESCO and Parthenon Publishing Group, Paris, France
7. Yadav PK (2013) Slash-and-burn agriculture in north-east India. J Expert Opin Environ Biol 2:1. https://doi.org/10.4172/2325-9655.1000102
8. Pandey DK, Adhiguru P, Devi, SV, Dobha S. Dubey SK, Mehra TS (2019) Quantitative assessment of crop species diversity in shifting cultivation system of Eastern Himalaya. Curr Sci 117:1357–1363. https://doi.org/10.18520/cs/v117/i8/1357-1363
9. Pandey DK, Adhiguru P, Momin KC, Kumar P (2022) Agrobiodiversity and agroecological practices in 'jhumscape' of the Eastern Himalayas: don't throw the baby out with the bathwater. Biodivers Conserv 31:2349–2372. https://doi.org/10.1007/s10531-022-02440-7
10. Hazarika TK (2012) Citrus genetic diversity of north-east India, their distribution, ecogeography and ecobiology. Genet Resour Crop Evol 59:1267–1280. https://doi.org/10.1007/s10722-012-9846-2
11. Das B, Sengupta S, Parida SK, Roy B, Ghosh M, Prasad M, Ghose TK (2013) Genetic diversity and population structure of rice landraces from Eastern and North Eastern States of India. BMC Genet 14:71. https://doi.org/10.1186/1471-2156-14-71
12. Singh AR, Devi EL, Dayal V, Saha S, Lungmauna, Dutta SK, Boopathi T, Singh SB (2019) Diversity of landraces maize in Mizoram: prospect, challenges and opportunities. In: Saha S, Lungmuana, Dayal V, Ansari MA, Jat SL, Chowdhury S, Lalliansanga S, Shakuntala I, Saithantluanga H, Rakshit S, Prakash N (eds) Souvenir: National workshop on scientific maize cultivation in North East India, March 05 2019, Aizawl, Mizoram, ICAR Research Complex for NEH Region, Umiam, Meghalaya, India, pp 98–104
13. Borggarrd O, Gafur A, Petersen L (2003) Sustainability appraisal of shifting cultivation in the Chittagong Hill Tracts of Bangladesh. Ambio 32:118–123. https://doi.org/10.1579/0044-7447-32.2.118
14. Ray SK, Chatterjee D, Chowdhury P, Deka BC, Bihari P, Saha S (2021) Comparison of indigenous and mechanical conservation technologies for shifting cultivation agro-ecology of north-eastern Himalaya. J Indian Soc Soil Sci 69:60–71. https://doi.org/10.5958/0974-0228.2021.00021.9
15. Neli V (2022) Terrace rice fields: a cultural heritage of the Angami Nagas. In: Chophy GK, Chaudhuri SK (eds) The cultural heritage of Nagaland. Routledge, London UK, pp 315–330
16. Cairns M, Keitzar, Yaden TA (2007) Shifting forests in Northeast India: management of Alnus nepalensis as an improved fallow in Nagaland. In: Cairns M (ed) Voices from the forest: integrating indigenous knowledge into sustainable upland farming. Routledge, Abingdon, Oxon, UK, pp 341–378
17. Giri K, Mishra G, Jayara RSC, Kumar R (2018) Agro-bio-cultural diversity of alder based shifting cultivation practiced by Angami tribes in Khonoma Village, Kohima, Nagaland. Curr Sci 115:598–599

18. Singh RA, Gupta RC (2002) Traditional land and water management systems of North-East hill region. Indian J Tradit Knowl 1:32–39
19. Kalidas-Singh S, Bora IP, Thakuria, D, Changkija S, Puyam A, Khwairakpam L (2021) The comparative soil fertility in traditional alder-based shifting cultivation of varied fallow lengths in Eastern Indian Himalayas. Soil Sci Plant Nutr 67:681–692. https://doi.org/10.1080/00380768.2021.2009741
20. Rathore SS, Karunakaran K, Prakash B (2010) Alder based farming system a traditional farming practice in Nagaland for amelioration of *jhum* land. Indian J Tradit Knowl 9:677–680
21. Changkija S (2015) Sustainable indigenous agroforestry farming system based on Alder trees (*Alnus nepalensis*) practiced by Naga tribes in Nagaland, India. In: Eckman K, Ralte L (eds) Integrated land use management in the Eastern Himalayas. Akansha Publishing House, New Delhi, India, pp 98–113
22. Sharma E, Ambasht RS (1984) Seasonal variation in nitrogen fixation by different ages of root nodules of *Alnus nepalensis* plantations, in the eastern Himalayas. J Appl Ecol 21:265–270. https://doi.org/10.2307/2403052
23. Sharma E, Ambasht RS (1986) Nitrogen accretion and its energetics in the Himalayan alder. Funct Ecol 2:229–235. https://doi.org/10.2307/2389699
24. Giri K (2019) The first report of indigenous free-living diazotroph *Kosakonia sacchari* isolated from Himalayan alder-based shifting cultivation system in Nagaland, India. J Soil Sci Plant Nutr 19:574–579. https://doi.org/10.1007/s42729-019-00056-5
25. Perakis SS, Pett-Ridge JC (2019) Nitrogen-fixing red alder trees tap rock-derived nutrients. Proc Natl Acad Sci USA 116:5009–5014. https://doi.org/10.1073/pnas.1814782116
26. Mishra G, Giri K, Pandey S (2018) Role of *Alnus nepalensis* in restoring soil fertility: a case study in Mokokchung Nagaland. Natl Acad Sci Lett 41:265–268. https://doi.org/10.1007/s40009-018-0668-4
27. Murray N, Das S (2021) Zabo farming system—a sustainable farming method base on traditional knowledge for natural resource management practices by tribal in Nagaland, India. Int J Agric Environ Biotechnol 14:203–205. https://doi.org/10.30954/0974-1712.02.2021.13
28. Singh RK, Hannah K, Asangla BR, Borkotoky R (2018) Zabo: a time-tested integrated farming system practiced by Chakhesang tribe of Nagaland. Indian J Hill Farming 31:188–192
29. Amenla I, Shuya K (2021) Zabo (zabü) farming of Kikruma village, Nagaland, India. In: Anon (ed) Innovations in agricultural extension, Michigan State University Extension, East Lansing, MI, USA & National Institute of Agricultural Extension Management, Hyderabad, India, pp 3–1–3–11
30. Kumar R, Patra MK, Thirugnanavel A, Deka BC, Chatterjee D, Borah TR, Rajesha G, Talang HD, Ray SK, Kumar M, Upadhyay PK (2018) Comparative evaluation of different integrated farming system models for small and marginal farmers under the Eastern Himalayas. Indian J Agric Sci 88:1722–1729. https://doi.org/10.56093/ijas.v88i11.84913
31. Das A, Layek J, Ramakrushna GI, Babu S, Devi MT, Dey, U, Suting D, Yadav GS, Lyngdoh DBD, Prakash N (2019) Integrated organic farming system: an innovative approach for enhancing productivity and income of farmers in North eastern hill region of India. Indian J Agric Sci 89:1267–1272. https://doi.org/10.56093/ijas.v89i8.92843
32. Chatterjee D, Kuotsu R, Ray SK, Patra MK, Thirugnanavel A, Kumar R, Borah TR, Chowdhury P, Pongen I, Satapathy BS, Deka BC (2022) Preventing soil degradation in shifting cultivation using integrated farming system models. Arch Agron Soil Sci 68:1841–1857. https://doi.org/10.1080/03650340.2021.1937139
33. Hombegowda HC, Jakhar P, Madhu M, Marwein Y (2021) 'Thang bun': indigenous practice of *in situ* biochar preparation-cum-application for improved jhum cultivation in North East India. Curr Sci 120:1160–1168
34. Plaimart J, Acharya K, Mrozik W, Davenport RJ, Vinitnantharat S, Werner D (2021) Coconut husk biochar amendment enhances nutrient retention by suppressing nitrification in agricultural soil following anaerobic digestate application. Environ Pollut 268:115684. https://doi.org/10.1016/j.envpol.2020.115684

35. Dubey SK, Sah U (2009) Indigenous nur bun method of potato cultivation in Meghalaya Hills. Asian Agri Hist 13:147–153
36. Prokop P, Kruczkowska B, Syiemlieh HJ, Bucała-Hrabia A (2017) Impact of topography and sedentary swidden cultivation on soils in the hilly uplands of North East India. Land Degrad Dev 29:2760–2770. https://doi.org/10.1002/ldr.3018
37. Jeffery S, Verheijen FGA, van der Veide M, Bastos AC (2011) A quantitative review of the effects of biochar application to soils on crop productivity using meta-analysis. Agric Ecosyst Environ 144:175–187. https://doi.org/10.1016/j.agee.2011.08.015
38. Abiven S, Schmidt MWI, Lehmann J (2014) Biochar by design. Nature Geosci 7:326–327. https://doi.org/10.1038/ngeo2154
39. Tiwari BK (2022) Temporal changes and innovation in raised-bed shifting cultivation at higher elevations in Meghalaya, Northeast India. In: Cairns M (ed) Farmer innovation and best practices by shifting cultivators in Asia and Pacific. CAB International, Wallingford, UK
40. Upadhaya K, Barik SK, Kharbhih VM, Nongbri G, Debnath G, Gupta A, Ojha A (2020) Traditional bun shifting cultivation practice in Meghalaya, Northeast India. Energy Ecol Environ 5:34–46. https://doi.org/10.1007/s40974-019-00144-3
41. Kumar A, Ramakrishnan PS (1990) Energy flow through an Apatani village ecosystem of Arunachal Pradesh in Northeast India. Hum Ecol 18:315–336. https://doi.org/10.1007/BF00889158
42. Saika SK, Das DN (2004) '*Aji gnui assonii*'—a practice of organic hill farming among the Apatani tribe of Eastern Himalaya. Int J Sustain Dev World Ecol 11:211–217. https://doi.org/10.1080/13504500409469825
43. Rai SC (2005) Apatani paddy-cum-fish cultivation: an indigenous hill farming system of North East India. Indian J Tradit Knowl 4:65–71
44. Saikia SK, Das DN (2008) Rice-fish culture and its potential in rural development: a lesson from Apatani farmers, Arunachal Pradesh, India. J Agric Rural Dev 6:125–131
45. Saika SK, Das DN (2015) Sustainable aquaculture: agro-ecological role of periphyton in ricefish farming. Rev Aquac 7:172–186. https://doi.org/10.1111/raq.12062
46. Tangjang S, Nair PKR (2015) Rice + fish farming in homesteads: sustainable natural-resource management for subsistence in Arunachal Pradesh, India. J Environ Sci Eng 4:545–557
47. Hussain SM, Debnath P, Sen D, Pathak M, Nabam J, Baruah D (2018) Integrated rice fish farming system in Arunachal Pradesh: an over view. Indian J Hill Farming 31:52–56
48. Baruah D, Posti R, Kunal K, Ganie PA, Tandel RS, Sarma D, Garima TK, Rinyo G (2019) Integrated rice-fish farming in hilly terraces of the Apatani Plateau, Arunachal Pradesh. Aquac Asia 23(2):24–35
49. Xie J, Hu L, Tang J, Wu X, Li N, Yuan Y, Yang H, Zhang J, Luo S, Chen X (2011) Ecological mechanisms underlying the sustainability of the agricultural heritage rice-fish coculture system. Proc Natl Acad Sci USA 108:E1381–E1387. https://doi.org/10.1073/pnas.1111043108
50. Deori BG (2020) Exploring nature-culture linkages in the Apatani Valley, Arunachal Pradesh, India. J World Herit Stud Spec Issue 17–23
51. Kanwat M, Bhagawati R, Singh M, Panwar AS, Kumar Singh AK (2009) Traditional wisdom of Apatanis: wet rice cultivation. In: Pawar AS, Ngachan SB, Munda GC, Das A (eds) Sustainable crop production for food sufficiency in N.E. India. ICAR Research Complex for NEH Region, Umiam, Meghalaya, India, pp 471–478
52. Farooqy P (2017) Symbiosis between nature & culture—a case study of the Apatani cultural landscape, India. J World Herit Stud Spec Issue 51–55
53. Nair PKR (1993) An introduction to agroforestry. Kluwer Academic Publishers, Dordrecht, The Netherlands
54. Kumar BM, Handa AK, Dhyani SK, Arunachalam A (2018) Agroforestry in the Indian Himalayan region: an overview. In: Gordon AM, Newman SM, Coleman BRW (eds) Temperate agroforestry systems, 2nd edn. CAB International, Wallingford, UK, pp 153–172. https://doi.org/10.1079/9781780644851.0153
55. Deb S (2020) Traditional agroforestry systems of Northeast India. In: Roy N, Roychoudhury S, Nautiyal S, Agarwal S, Baksi S (eds) Socio-economic and eco-biological dimensions in

resource use and conservation: strategies for sustainability. Springer Nature, Switzerland, pp 103–115. https://doi.org/10.1007/978-3-030-32463-6_5
56. Saikia P, Khan ML (2014) Homegardens of upper Assam, northeast India: a typical example of on-farm conservation of Agarwood (*Aquilaria malaccensis* Lam.). Int J Biodivers Sci Ecosyst Serv Manage 10:262–269. https://doi.org/10.1080/21513732.2014.973449
57. Sharma E, Sharma R, Sharma G, Rai SC, Sharma P, Chettri N (2008) Values and services of nitrogen-fixing alder based cardamom agroforestry systems in the Eastern Himalayas. In: Snelder DJ, Lasco RD (eds) Smallholder tree growing for rural development and environmental services. Dordrecht, The Netherlands, pp 393–409. https://doi.org/10.1007/978-1-4020-8261-0_1
58. Mathur I, Bhattacharya P (2022) Transition from shifting cultivation to agroforestry: a case study of regrouped villages in Tripura, India. Environ Chall 7:10047. https://doi.org/10.1016/j.envc.2022.100471
59. Reang D, Sahoo UK, Giri K, Hazarika A, Nath AJ (2022) Ethnic homestead forests of North-East India revealed as diverse land-use systems. Agroforest Syst 96:465–478. https://doi.org/10.1007/s10457-021-00700-z
60. Tangjang S, Nair PKR (2016) Integrated bamboo + pine homegardens: a unique agroforestry system in Ziro valley of Arunachal Pradesh, India. Int J Environ Agric Res 2:25–34
61. Ahirwal J, Sahoo UK, Thangjam U, Thong P (2022) Oil palm agroforestry enhances crop yield and ecosystem carbon stock in northeast India: implications for the United Nations sustainable development goals. Sustain Prod Consum 30:478–487. https://doi.org/10.1016/j.spc.2021.12.022
62. Sarangi SK, De LC, Bagra G, Nyori I (2006) Tokopalm and cane: potential non-timber agroforestry tree-crops of Arunachal Pradesh, In: Bhatt BP, Bujarbaruah KM (eds) Agroforestry in North East India: opportunities and challenges. ICAR Complex for NEH Region, Umiam, Meghalaya, India, pp 237–247
63. Kumar BK, Nair PKR (2006) Tropical homegardens: a time-tested example of sustainable agroforestry. Springer, Dordrecht, The Netherlands
64. Sahoo UK (2009) Traditional home gardens and livelihood security in North-East India. J Food Agric Environ 7:665–670
65. Barbhuiya AR, Sahoo UK, Upadhyaya K (2016) Plant diversity in the indigenous home gardens in the eastern Himalayan region of Mizoram, Northeast India. Econ Bot 70:115–131. https://doi.org/10.1007/s12231-016-9349-8
66. Stevens CJ (2019) Nitrogen in the environment. Science 363:578–580. https://doi.org/10.1126/science.aav8215
67. Eernstman N, Wals AEJ (2009) Jhum meets IFOAM: introducing organic agriculture in a tribal society. Int J Agric Sustain 7:95–106. https://doi.org/10.3763/ijas.2009.0398
68. Pangging G, Arunachalam A (2008) Is certification needed for 'jhum' cultivation? Curr Sci 95:1655–1656
69. Singh DP, Prakash HG, Swapna M, Solomon S (eds) (2023) Organic crop production management: focus on India, with global implications. Apple Academic Press, Palm Bay, FL, USA
70. Koohafkan P, Altieri MA (2011) Globally important agricultural heritage systems: a legacy for the future. Food and Agriculture Organization of the United Nations, Rome, Italy
71. Bhalerao AK, Rasche L, Scheffran J, Schneider UA (2022) Sustainable agriculture in Northeastern India: how do tribal farmers perceive and respond to climate change? Int J Sustain Dev World Ecol 29:291–302. https://doi.org/10.1080/13504509.2021.1986750
72. Zahan Y, Mahanta R, Rajesh PV, Goswami BN (2021) Impact of climate change on North-East India (NEI) summer monsoon rainfall. Clim Change 164:2. https://doi.org/10.1007/s10584-021-02994-5
73. Altieri MA, Nicholls CI (2017) The adaptation and mitigation potential of traditional agriculture in a changing climate. Clim Change 140:33–45. https://doi.org/10.1007/s10584-013-0909-y
74. Cui J, Liu H, Wang H, Wu S, Bashir MA, Reis S, Sun Q, Xu J, Gu B (2023) Rice-animal co-culture systems benefit global sustainable intensification. Earth Future 11:e2022EF002984. https://doi.org/10.1029/2022EF002984

75. Isbell F, Adler PR, Eisenhauer N, Fornara D, Kimmel K, Kremen C, Letourneau DK, Liebman M, Polley HW, Quijas S, Scherer-Lorenzen M (2017) Benefits of increasing plant diversity in sustainable agroecosystems. J Ecol 105:871–879. https://doi.org/10.1111/1365-2745.12789
76. Thapa R, Chatterjee A, Awale R, McGranahan DA, Daigh A (2016) Effect of enhanced efficiency fertilizers on nitrous oxide emissions and crop yields: a meta-analysis. Soil Sci Soc Am J 80:1121–1134. https://doi.org/10.2136/sssaj2016.06.0179
77. Bongers F, Chazdon R, Poorter L, Peña-Claros M (2015) The potential of secondary forests. Science 348:642–643. https://doi.org/10.1126/science.348.6235.642-c
78. Layek J, Narzari R, Hazarika S, Das A, Rangappa K, Devi S, Balusamy A, Saha S, Mandal S, Idapuganti RG, Babu S, Choudhury BU, Mishra VK (2022) Prospects of biochar for sustainable agriculture and carbon sequestration: an overview for Eastern Himalayas. Sustainability 14:6684. https://doi.org/10.3390/su14116684
79. Antonelli A (2023) Indigenous knowledge in key to sustainable food systems. Nature 613:239–242. https://doi.org/10.1038/d41586-023-00021-4

Agrobiodiversity and Crop Productivity of the Alder-Based Jhum Farming System in Khonoma Village, Nagaland, India

Krishna Giri, Gaurav Mishra, R. S. C. Jayaraj, and Meribeni Tsopoe

Abstract Alder-based jhum cultivation practiced by the Angami tribe in Khonoma, Nagaland is a traditional farming system that conserves soil nutrients and maintains agrobiodiversity. A study was conducted to assess agrobiodiversity, major soil nutrients and crop productivity of this farming system. On-farm observations found a total of 14 crops grown by the farmers; crop diversity was dominated by vegetables followed by maize, spices, pulses, and millets. Potato was the most common vegetable crop in this farming system grown by almost all the farmers, followed by cabbage. Soil nutrient analyses found 635 ± 14.9, 32.7 ± 2.4 and 30.8 ± 3.1 mg kg^{-1} available nitrogen, phosphorus and potassium, respectively. In addition, $3.81 \pm 0.15\%$ organic carbon was determined to be present in the soil. Among the cultivated vegetables, the yield of dominant field crops i.e., potato, cabbage, green leafy vegetables (lai patta) and coriander leaf was estimated to be 9.24, 15.2, 5.0 and 5.67 Mg ha^{-1} respectively in the year 2016, while productivity in 2017 was estimated to be 12.3, 11.3, 9.0 and 6.0 Mg ha^{-1}, respectively. Due to this unique form of agriculture and rich natural biodiversity, the village has been recognised as *"the first green village of Asia"* and a tourist destination in Nagaland.

Keywords Agrobiodiversity · Alder jhum farming · Crop productivity · Khonoma village · Soil nutrients

K. Giri (✉) · G. Mishra · R. S. C. Jayaraj · M. Tsopoe
ICFRE-Rain Forest Research Institute, Jorhat, Assam 785010, India
e-mail: krishna.goswami87@gmail.com

K. Giri · G. Mishra
Centre of Excellence on Sustainable Land Management, Indian Council of Forestry Research and Education, Dehradun, Uttarakhand 428006, India

R. S. C. Jayaraj
Ex-Principal Chief Conservator of Forests, Andaman and Nicobar Islands 744101, India

S. K. Tripathi and F. Q. Brearley (eds.), *Shifting Cultivation Systems*,
https://doi.org/10.1007/978-3-031-70388-1_5

1 Introduction

Jhum cultivation system (i.e. shifting cultivation) in North-east India is believed to have originated in the Neolithic period around 7000 B.C. [1]. Due to mountainous terrain, settled cultivation constitutes only a small portion of arable land in the river valleys, which is popularly known as wet terrace cultivation. In the entire region, diverse and tribe-specific models of jhum cultivation blended with unique traditional innovations are prevalent [2]. Among these traditional jhum farming systems, Himalayan alder-based agroforestry cum jhum cultivation practice of the Angami tribe in Khonoma Village, Nagaland, is one of the more well-known farming systems in the eastern Himalayas. Alder (*Alnus nepalensis*; Betulaceae) is a fast-growing multipurpose tree species found at an altitudinal range of 1000–2000 m above sea level in North-east India. In Khonoma village, it is known as Rupuo and mature trees are pollarded 2–3 m from the ground. The wood of pollarded trees is used for shade, fodder, fuel wood, timber and charcoal making, while twigs, small branches and leaves are burnt in the field before cropping in a jhum cycle [3]. The land is cropped systematically in a cyclical manner for two to three years and left fallow for the next five to seven years (Fig. 1). During the fallow period, the farmers shift to other sites for crop cultivation and allow the pollarded trees to re-establish. The leaf litter in the fallow period contributes large quantities of organic matter to the soil and the tree roots enrich the soil through symbiotic nitrogen fixation in association with *Frankia* bacteria [4]. This unique jhum cultivation-cum-agroforestry practice is the cultural heritage of Khonoma village since its origin [5]. The community grows vegetables, maize (*Zea mays*; Poaceae), spices, pulses and Job's tears (*Coix lacryma-jobi*; Poaceae) in this jhum farming system dominated by potato (*Solanum tuberosum*; Solanaceae) and cabbage (*Brassica oleracea*; Brassicaceae). The vegetable crops supplement the staple paddy (*Oryza sativa*; Poaceae) crop grown in the wet terrace farm (pani kheti) down-slope from the village. The soil nutrients from alder-grown areas are carried to the wet terrace farm via the stream and enrich the soil fertility and crop productivity of the terrace fields. Past studies in this village have focussed on soil fertility and traditional management practices of the Angami tribe [5–9]. The present study was undertaken with the objective to assess farm agrobiodiversity and productivity of this unique alder-based jhum cultivation system.

2 Methods

The study was conducted in Khonoma Village, Kohima, Nagaland. Field visits were carried out to record cultivated plant diversity in the alder-based jhum cultivation system during the crop growing phase, i.e. March to October 2016 and 2017. The number of crops and vegetables grown in alder-based jhum fields were recorded. In addition, soil samples (0–30 cm depth) were collected from the fields using a simple random sampling method and brought to the laboratory. Samples were processed for

Fig. 1 **a** Alder-based jhum fallow and **b** reestablishment of pollarded alder trees in Khonoma village, Nagaland, North-east India

estimation of pH, available nitrogen and exchangeable potassium using the methods of Jackson [10], phosphorus using Bray and Kurtz [11], and organic carbon following Walkley and Black [12]. The major vegetable crops viz. potato, cabbage, lai patta (*Brassica juncea*; Brassicaceae) and coriander (*Coriandrum sativum*; Apiaceae) were harvested using 1 m × 1 m plots. The edible parts of the harvested crops within the plots were collected and the fresh weight was recorded for yield estimation. The data presented is the mean of ten replicate representing the entire alder-grown jhum area in Khonoma village.

3 Results and Discussion

The present study assessed crop diversity of alder-based jhum farming/agroforestry system developed by the Angami community in Khonoma Village, Nagaland. Field observations found a total of 14 crops consisting of vegetables, maize, legume pulses, spices and millets (Table 1). Cultivation of 14 crops in a given jhum field provides multiple nutrients to the farming communities and sustains their livelihoods. Ramakrishnan [13], while speaking in defence of shifting cultivation, mentioned that nutrients are released in a single flush through slash-and-burn. The resultant nutrients are leached and lost rapidly because of the steep slopes and heavy rainfall. So the farmers cultivate a large variety of crops: at times as many as 30–35 crops to make the best use of the nutrients [14]. Through a diverse cropping system, farmers simultaneously ensure protein, cereal and fibre requirements from a single field. The alder-based practice is mainly focused on vegetables and some cash crop cultivation such as garlic (*Allium sativum*; Amaryllidaceae), ginger (*Zingiber officinale*; Zingiberaceae), spices, oil seeds, chillies etc.

Potato was the dominant crop in the alder-based jhum farming system followed by cabbage, maize and millet. Additionally, vegetables from the Cucurbitaceae family, soyabean (*Glycine max*; Leguminosae), French bean (*Phaseolus vulgaris*; Leguminosae) and oil seed (*Perilla frutescens*; Lamiaceae) were found cultivated in the mixed cropping system (Fig. 2).

In previous studies, crop diversity in jhum farming systems has also been reported from 15 to 60 crops in a single field [2, 14]. In traditional slash-and-burn cultivation systems, paddy, maize and millet are grown as staple crops while a variety of vegetables, tuber crops, cereals, legume pulses, oilseed crops and green leafy vegetables are grown as secondary or mixed crops. The observations of the present study are also in conformity with the crop diversity reports in jhum farming systems [13, 14]. However, vegetables are the primary food crops in the alder-based jhum cultivation system.

Table 1 Crop species richness in an alder-based jhum farming system of Khonoma village, Nagaland, North-east India

Common name	Angami name	Latin name	Family
Tomato	Bengano	*Solanum lycopersicum*	Solanaceae
Potato	Aloo	*Solanum tuberosum*	Solanaceae
Chilli	Chiisi	*Capsicum annuum*	Solanaceae
Maize	Shiiko	*Zea mays*	Poaceae
Cabbage	Kobi	*Brassica oleracea*	Brassicaceae
Job's tears (Millett)	Kerhi	*Coix lacryma-jobi*	Poaceae
Squash	Eskus	*Sechium edule*	Cucurbitaceae
Cucumber	Tsiituo	*Cucumis sativus*	Cucurbitaceae
Ginger	Kevii	*Zingiber officinale*	Zingiberaceae
Garlic	Chiimerie	*Allium sativum*	Amaryllidaceae
Soyabean	Shiidza	*Glycine max*	Leguminosae
French bean	Riicii	*Phaseolus vulgaris*	Leguminosae
Oil seed	Tier-sii	*Perilla frutescens*	Lamiaceae
Green leafy vegetables (Lai patta)	Gakrie	*Brassica juncea*	Brassicaceae

Fig. 2 Potato cultivation in Khonoma village, Nagaland, North-east India

Soil analyses showed 635 ± s.e. 14.9, 32.7 ± 2.43 and 30.8 ± 3.07 mg kg^{-1} available nitrogen, phosphorus and potassium, respectively. In addition, 3.81 ± 0.15% organic carbon was determined in the soil (Table 2). Soil organic carbon in alder-based jhum land was estimated to be higher, nitrogen and phosphorus in the medium range, and exchangeable potassium lower than the soil nutrient ranges fixed for agricultural soils by other scientists working in the region [15, 16]. Soil pH was observed in the acidic range, which will be due to the abundance of water-soluble ions of iron and aluminium in North-east Indian soils. Chase and Singh [9] also studied soil nutrients and fertility of natural forest, alder-based jhum fallow and wet terrace paddy cultivation system in Khonoma and reported a soil pH of 4.82, 160 kg ha^{-1} available nitrogen, 17.2 kg ha^{-1} phosphorus, 131 kg ha^{-1} potassium and 2.37% soil organic carbon in alder-based jhum fields. The order of soil fertility across the studied land-use systems studied by them was: natural forest > alder-based jhum > paddy field.

This study also estimated productivity of four dominant crops i.e. potato, cabbage, coriander and lai patta leaves in the alder-based farming system (Table 3) which is the first attempt to estimate the crop yield in such a farming system. The productivity of potato was greater than the district average of 5.87 Mg ha^{-1} in Kohima district [17]. However, productivity of cabbage was lower than the district and state average i.e. 19.5 and 20.0 Mg ha^{-1} respectively [18]. Similarly the productivity of alder-cardamom (*Amomum subulatum*; Zingiberaceae) agroforestry systems in Sikkim is

Table 2 Soil nutrients in alder-based jhum cultivation land in Khonoma Village, Nagaland, North-east India (values are mean ± standard error, $n = 10$)

Soil parameter	Value
pH	4.10 ± 0.05
$N_{avail.}$ (mg kg^{-1})	635 ± 14.9
$P_{avail.}$ (mg kg^{-1})	32.7 ± 2.43
$K_{exch.}$ (mg kg^{-1})	30.8 ± 3.07
SOC (%)	3.81 ± 0.15

Table 3 Productivity of selected crops in alder-based jhum fields in Khonoma village, Nagaland, North-east India over two years (values are mean ± standard error, $n = 10$)

Crop	Latin name	Productivity (Mg ha^{-1})	
		2016	2017
Cabbage	*Brassica oleracea*	15.2 ± 0.95	12.3 ± 1.23
Potato	*Solanum tuberosum*	9.24 ± 0.11	11.3 ± 1.20
Lai patta	*Brassica juncea*	5.00 ± 0.12	8.96 ± 0.98
Coriander	*Coriandrum sativum*	5.67 ± 0.43	5.99 ± 0.56

also very high in young plantations ranging from 250 to 350 kg ha^{-1} as compared to other cropping systems in the region [19].

In conclusion, it was observed that the alder-based farming system developed by the Angami tribe in Khonoma village is able to sustain the food, fuel wood and other requirements of the villagers on one hand, and preserve natural biodiversity and carbon stocks on the other. This farming system produces multiple nutritious crops on hill slopes of the village supplementing the staple paddy crop grown in the wet terrace.

Acknowledgements The authors are thankful to the farmers of Khonoma Village for their assistance during the study. Research facilities provided by ICFRE-Rain Forest Research Institute, Jorhat, Assam and financial assistance received from the Indian Council of Forestry Research and Education, Dehradun are gratefully acknowledged.

References

1. Borthakur DN (2004) Cropping systems. In: Singh VP, Sharma N, Ojha CSP (eds) The Brahmaputra basin water resources. Springer, Dordrecht, The Netherlands, pp 401–410. https://doi.org/10.1007/978-94-017-0540-0_22
2. Nagaland Environmental Protection and Economic Development and International Institute of Rural Reconstruction (1999) Building upon traditional agriculture in Nagaland, India. Nagaland Environmental Protection and Economic Development, Nagaland, India and International Institute of Rural Reconstruction, Silang, Cavite, The Philippines
3. Rana SK, Rana HK, Shreshta KK, Sujakhu S, Ranjitkar S (2018) Determining bioclimatic space of Himalayan alder for agroforestry systems in Nepal. Plant Divers 40:1–18. https://doi.org/10.1016/j.pld.2017.11.002
4. Khan A, Myrold DD, Mishra AK (2007) Distribution of Frankia genotypes occupying *Alnus nepalensis* nodules with respect to altitude and soil characteristics in the Sikkim Himalayas. Physiol Plant 130:364–371. https://doi.org/10.1111/j.1399-3054.2006.00872.x
5. Giri K, Mishra G, Jayaraj RSC, Kumar R (2018) Agrobio-cultural diversity of alder based shifting cultivation practiced by Angami tribes in Khonoma Village, Kohima, Nagaland. Curr Sci 115:598–599
6. Giri K (2019) The first report of indigenous free-living diazotroph *Kosakonia sacchari* isolated from Himalayan alder-based shifting cultivation system in Nagaland, India. J Soil Sci Plant Nutr 19:574–579. https://doi.org/10.1007/s42729-019-00056-5

7. Mishra AK, Sharma UC (2001) Traditional wisdom in range management for resource and environment conservation in North Eastern region of India. ENVIS Bull Himal Ecol 9(1):31–36
8. Rathore SS, Karunakaran K, Prakash B (2010) Alder based farming system: a traditional farming practices in Nagaland for amelioration of *jhum* land. Indian J Tradit Knowl 9:677–680
9. Chase P, Singh OP (2014) Soil nutrients and fertility in three traditional land use systems of Khonoma, Nagaland, India. Resour Environ 4:181–189. https://doi.org/10.5923/j.re.20140404.01
10. Jackson ML (1973) Soil chemical analysis, 2nd edn. Prentice Hall of India Pvt. Ltd., New Delhi, India
11. Bray RH, Kurtz LT (1945) Determination of total, organic, and available forms of phosphorus in soils. Soil Sci 59:39–46
12. Walkley A, Black IA (1934) An examination of the Degtjareff method for determining soil organic matter and a proposed modification of the chromic acid titration method. Soil Sci 37:29–38
13. Ramakrishnan PS (1993) In defence of shifting agriculture. https://www.downtoearth.org.in/interviews/in-defence-of-shifting-agriculture-31654. Accessed: 4 July 2015
14. Pandey DK, Adhiguru P, Momin KC, Kumar P (2022) Agrobiodiversity and agroecological practices in 'jhumscape' of the Eastern Himalayas: don't throw the baby out with the bathwater. Biodivers Conserv 31:2349–2372. https://doi.org/10.1007/s10531-022-02440-7
15. Purkayastha A, Varghese S, Purkayastha A (2011) Nutrient indices of Barak Valley tea soils with reference to organic matter and soil available potash status. Two Bud 58:118–119
16. Department of Agriculture & Cooperation, Ministry of Agriculture, Government of India (2012) Annual report 2011–2012. Department of Agriculture & Cooperation, Ministry of Agriculture, Government of India, New Delhi, India
17. Venyo V, Sharma A (2018) Economics and constraint of potato crop in Kohima district of Nagaland. Technofame J Multidiscip Adv Res 7:75–80
18. Koza A, Sahu AK, Das S, Patra NK (2018) Economics of cabbage cultivation and post-harvest losses in Phek district of Nagaland. Econ Aff 63:229–235. https://doi.org/10.30954/0424-2513.2018.00150.28
19. Sharma G, Sharma R, Sharma E (2009) Traditional knowledge systems in large cardamom farming: biophysical and management diversity in Indian mountainous regions. Indian J Tradit Knowl 8:17–22

Linking Soil Organic Carbon and Soil Parameters in Shifting Cultivation Sites of Nagaland (North-East India)

Gaurav Mishra, Abhishek Jangir, and Rosa Francaviglia

Abstract Shifting cultivation is one of the oldest and most common agricultural practices in hilly states of North-east India, and a central element of land management and livelihoods for local people. In addition to maintaining agricultural biodiversity, shifting cultivation can influence soil health by driving soil nutrient losses and depleting soil organic carbon (SOC). Assessment of these ecosystem services is therefore important to determine the agronomic potential of cultivated soils. The present study was conducted in five soil profiles to assess soil properties and their relationship with SOC in selected shifting cultivation sites of Nagaland. The studied area was mainly underlain by Inceptisols and Entisols (suborders Humic Dystrudepts, Typic Dystrudepts, and Typic Udorthents) and soil texture varied across the different sites from clay to sandy clay loam, while the mean SOC content was 1.07% (range 0.32–2.67%) and decreased with soil depth. Soil organic carbon was significantly correlated with clay, clay ratio, total nitrogen, and cation exchange capacity. Since SOC, an indicator of overall soil quality and related to soil management, has chromogen properties, soil color attributes were also significantly correlated with SOC content (the darker the color the greater the SOC, and vice versa). This present study will help in providing information about the spatial variability of SOC in the eastern Himalayan region, which needs to be more precisely estimated, and provide insights into sustainable soil management in the areas where shifting cultivation is practiced.

Keywords Nagaland · Nutrients · Organic carbon · Shifting cultivation · Soil properties

---

G. Mishra
Centre of Excellence on Sustainable Land Management, Indian Council of Forestry Research and Education, Dehradun, Uttarakhand 248006, India

A. Jangir
National Bureau of Soil Survey and Land Use Planning, Regional Center, Indian Council of Agricultural Research, Udaipur, Rajasthan 313001, India

R. Francaviglia (✉)
Council for Agricultural Research and Economics, Research Centre for Agriculture and Environment, 00184 Rome, Italy
e-mail: r.francaviglia@gmail.com

S. K. Tripathi and F. Q. Brearley (eds.), *Shifting Cultivation Systems*,
https://doi.org/10.1007/978-3-031-70388-1_6

1 Introduction

Shifting cultivation is one of the major land-use systems of Nagaland, a state in North-east India, as driven by the socio-economic needs of local people that mainly adopt shifting cultivation to grow crops for family consumption or to sell vegetables and fruits in the nearby markets, but they rarely choose large-scale systems such as cash crop plantations. Thus, shifting cultivation is the major source of livelihoods for the local people and is still widely practiced, probably because it is the most suitable system that can be adopted in response to the natural environment and the physiographic condition of the region [1, 2]. However, shifting cultivation results in forest loss, with associated soil and nutrient losses by erosion and decreases in forest and soil biodiversity [3, 4]. The spatial extent of soil degradation processes is influenced by many factors besides land management, including soil type, climate, soil organisms and topography [5]. However, studies have mainly addressed the conversion of native vegetation to shifting cultivation or its comparison with other land-uses along with the effect on soil physical, chemical and biological properties.

Soil organic carbon (SOC) is an indicator of overall soil quality that affects physical, chemical and biological properties and provides ecosystem services such as climate and water regulation, water purification, nutrient cycling, soil biodiversity, etc. [6]. In addition, soil management can have positive effects on SOC, crop yields, runoff and soil erosion control, water quality, biodiversity, and climate change mitigation through carbon sequestration [7]. Soil formation, and consequently soil type, is regulated by a set of forming processes [5] i.e., climate (temperature and precipitation), organisms (natural vegetation, land-use and management, and soil biota), topography (slope, elevation, landscape position), parent material (soil mineralogy, texture and fertility) and time (over time, climate and organisms act on the soil mediated by parent material and topography leading to soil horizonation). In general, SOC is greatest under cool humid conditions and lowest under warmer and drier climates, however SOC content is more affected by temperature than precipitation [8]. Research also reported a predominant influence of soil type on SOC, particularly associated with soil moisture regime and soil texture [9], which can range from 31 to 776 Mg C ha^{-1} in the upper 100 cm in Arenosols and Histosols, respectively [10].

So far, no studies have considered how soil type affects soil quality and the distribution with depth of soil parameters under shifting cultivation. In this chapter, we aimed to assess the effect soil type on SOC, the relationship between SOC and some soil parameters, and the correlation between edaphic variables and topographical features.

2 Materials and Methods

2.1 *Study Area*

The study was performed in Mokokchung district of Nagaland state (North-east India). Nagaland is a small state located in the north-eastern region of India and the topography of the area is dominated by hill ranges and valleys [3]. The state area is 16,579 km^2 (83.5% is covered by forests) and over 80% of the population lives in villages and depends on rural and natural resources. Mokokchung is one of the 11 districts of Nagaland, and has a sub-tropical humid climate with average annual temperature up to 24 °C, while annual rainfall (1600–2000 mm) is influenced by the south-west monsoon (from mid-May to the end of September). The district has a total area of 1615 km^2, is located at 25° 56′–27° 40′ N and 93°53′–94° 53′ E, with elevation up to 1650 m asl [11]. Shifting cultivation, or jhum, is the most common land-use practice across Nagaland [12] including Mokokchung district. About 70% of farmers in Mokokchung practice shifting cultivation and the present average fallow cycle, during which the jhum field is abandoned and vegetation naturally develops to a secondary forest, is 10 years [11].

2.2 *Soil Profiles and Laboratory Analysis*

A pedological survey was conducted on five soil profiles under shifting cultivation (Table 1; Fig. 1).

Soil profiles were classified following the USDA Soil Taxonomy [13]. The pedo-morphological characteristics of soil profiles [14], i.e., horizons, depth and boundary, color, texture, structure, consistency, porosity and visual estimation of size and quantity of roots were determined. Soil samples from the different soil layers (the pedological horizons) were manually collected with a spade, air-dried at room temperature to constant weight, ground and passed through a 2-mm sieve to exclude litter, roots and coarse particles. The hydrometer method was used for particle size determination [15] to evaluate clay, silt and sand contents and texture following the USDA methodology [16]. The ratios sand/silt and silt/clay were also calculated, as well as the clay ratio using the equation:

$$\text{Clay ratio} = \text{clay } (\%)/[\text{sand } (\%) + \text{silt } (\%)] \quad (1)$$

Moist hue and color (value and chroma) were estimated with Munsell Soil Color Charts [17]. Bulk density (BD) was determined by the core method on undisturbed soil samples [18]. Soil pH was determined in 1:2 soil–water suspensions (Systronics System 361 pH meter), and electrical conductivity (EC) in 1:2.5 soil–water suspensions (Systronics System 307 EC meter). Soil organic carbon (SOC), total N, available phosphorus (Av. P), and available nitrogen (Av. N) were determined following

Table 1 Description of the study sites where soil profiles were surveyed in Mokokchung district, Nagaland, North-east India

Profile	Latitude (N)	Longitude (E)	Slope	Elevation (m)	Soil classification	Land-use
P1	26° 21.609′	94° 53.685′	3°	490	Inceptisol: Humic Dystrudept	Jhum: cardamom (*Amomum* sp.; Zingiberaceae), paddy (*Oryza sativa*; Poaceae) and banana (*Musa* sp.; Musaceae)
P4	26° 14.760′	94° 31.890′	20°	774	Inceptisol: Humic Dystrudept	Jhum: paddy (*Oryza sativa*), bitter gourd (*Momordica charantia*; Cucurbitaceae), corn (*Zea mays;* Poaceae) and toko palm trees (*Livistona jenkinsiana*; Arecaceae)
P6	26° 33.714′	94° 40.786′	6°	614	Inceptisol: Humic Dystrudept	Jhum: paddy (*Oryza sativa*)
P8	26° 30.518′	94° 40.021′	20°	889	Inceptisol: Typic Dystrudept	Jhum (2nd year): paddy (*Oryza sativa*)
P12	26° 30.825′	94° 26.208′	3°	445	Entisol: Typic Udorthent	Jhum: paddy (*Oryza sativa*), banana (*Musa* sp.) and bitter gourd (*Momordica charantia*)

the Walkley–Black [19], Kjeldahl [20], Bray-Kurtz [21] and Subbiah and Asija [22] methods, respectively.

Soil organic carbon stocks were calculated with the following equation [23]:

$$\text{SOC stock} = \text{SOC } (\%) \times \text{BD} \times \text{D} \quad (2)$$

where SOC stock is expressed in Mg C ha^{-1}, SOC (%) is organic carbon concentration, BD is bulk density in Mg m^{-3} and D is sample depth in cm.

For the determination of exchangeable Na^+ and K^+, the soil samples were saturated with 1 N ammonium acetate (pH 7.0) and Na^+ and K^+ concentration in the leachate was measured on a flame photometer [24]. The Ca^{2+} and Mg^{2+} ions were determined by saturating the soil samples with 1 N sodium chloride solution [25] and analysing by atomic absorption spectrophotometry (Analytical Technologies Ltd). For cation exchange capacity (CEC), soil samples were saturated with 1 N sodium acetate (pH 8.2). After removal of excess sodium acetate by washing with distilled water and alcohol, the absorbed Na^+ was extracted by washing with 1 N ammonium acetate (pH 7.0) and the leachate was made up to a known volume. The Na^+ ions present in

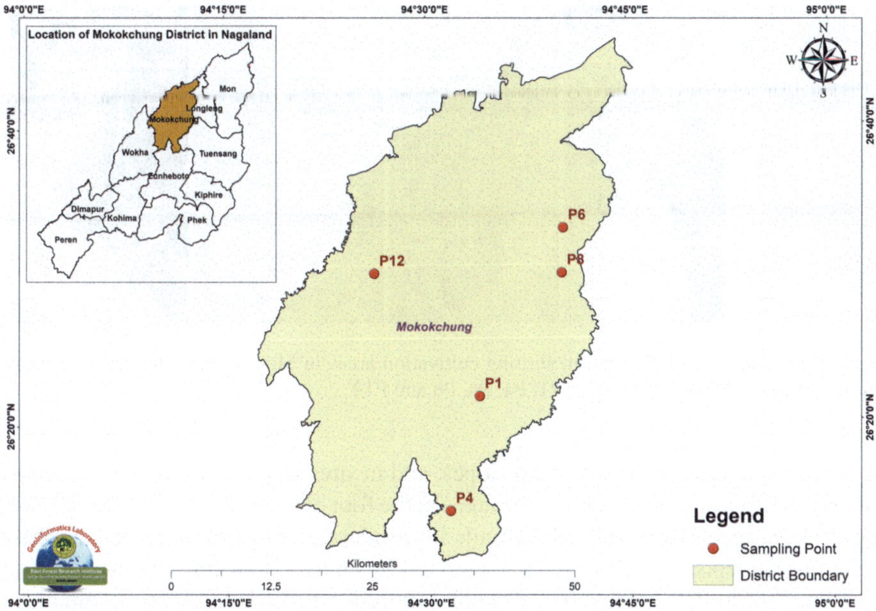

Fig. 1 Soil profile sampling points in Mokokchung district, Nagaland, North-east India

the leachate were determined with a Systronics flame emission spectrophotometer to calculate CEC [26].

2.3 *Statistical Analysis*

Descriptive statistics of soil parameters and correlations were performed with Statistica 8.0 (Statsoft, Tulsa, OK, USA). Spearman rank order correlations were based on 21 soil parameters, elevation and slope, highlighting significant correlations. The correlation matrix between SOC content and Munsell color attributes was based on moist value, chroma, value + chroma and SOC (%).

3 Results and Discussion

3.1 *Pedo-Morphological Characteristics*

Soils examined belonged to the Inceptisol and Entisol orders [13] that are both very young soils and not very developed (Fig. 2). Inceptisols (from Latin inceptum, 'beginning') are more developed than Entisols (from Latin recens or recentis, 'recent').

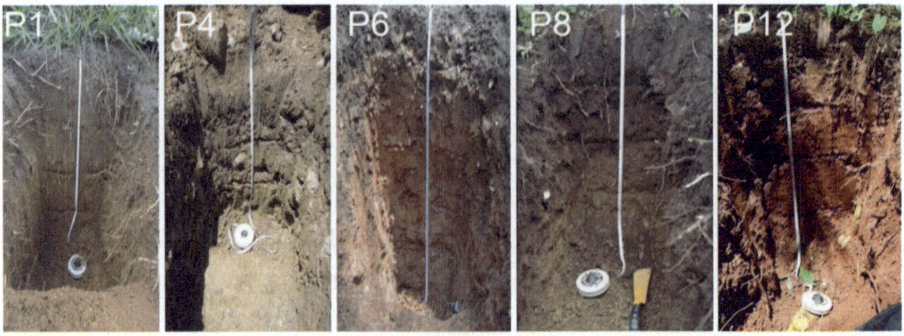

Fig. 2 Excavated soil profiles from shifting cultivation areas in Mokukchung district, Nagaland, North-east India. From left to right: P1, P4, P6, P8 and P12

These soils are typical of very steep slopes, and in sites where new non-pedogenised material is often supplied, such as the areas at the foot of steep slopes. Profiles P1, P4, and P6 were Inceptisols, suborder Humic Dystrudepts, and P8 belonged to suborder Typic Dystrudepts, and all were characterised by an ustic moisture regime and acid reaction. Profile P12 belonged to Entisols, suborder Typic Udorthents, common in humid temperate regions and developed on recent erosional surfaces.

Soils mainly differed in terms of depth and number of horizons, but also in color and texture. In particular, depth was limited to 90 cm in P1 and 125–150 cm in the other soil profiles. Similarities in soil horizons were found in P1 and P8 (A-Ap and Bw), P4 and P6 (A-Ap, Bw and BC), while P12 was quite different with no B horizon (A and C horizons only). In particular, the absence of a B horizon and the presence of a C horizon indicates that P12 was little affected by pedogenic processes. The presence of two BC horizons (transitional between B and C horizons) in P4 indicates that this soil was more developed than P6, while P1 and P8 had a similar development.

Hue (the basic color wavelength) was mainly 7.5 YR (with YR indicating yellow–red) with color from dark brown to strong brown in P4, and was 10 YR with color from yellowish brown to dark and very dark yellowish brown in the other profiles. Thus, P4 is less yellow and darker compared to the other profiles. The ranges of color values (darkness, scale 0–10) and chromas (strength, scale 0–8), were mainly 3–4 and 3–6, respectively, in P4 and were different compared with the other profiles, in agreement with hues (Table 2).

Darker soils (with lower value and chroma) have more SOC than lighter colored soils [27] and vice versa, thus color attributes (value, chroma and their sum) are correlated with SOC content (see Sect. 3.5).

Soil texture according to USDA [16]) classification was mainly medium (clay loam) in profiles P1 and P8, fine to medium (clay and clay loam) in P4, coarse to medium (sandy clay loam, clay loam and loam) in P6, and coarse (loamy sand, sand, and sandy loam) in P12 (Table 2). Thus, contrasting soil textures were found in P4 (medium-fine) and P12 (coarse).

Table 2 Pedo-morphological characteristics of soil profiles under shifting cultivation in Mokukchung district, Nagaland, North-east India

Horizon	Depth (cm)	Boundary[1]	Color hue, value and chroma (Moist)	Texture	Structure[2]	Consistency[3]			Porosity[4]	Roots[5]
ProfileP1: Inceptisols, humic dystrudepts		**D, T**			**G, S, T**	**Dry**	**Moist**	**Wet**	**S, Q**	**S, Q**
Ap	0–20	c s	10 YR 3/2	Sandy clay loam	WE, FI, SB	SHA	FR	SST and SPL	Few F	FM
Bw1	20–40	g s	10 YR 3/2	Clay loam	WE, FI, SB	HA	FR	SST and SPL	Few F	FM
Bw2	40–60	g s	10 YR 5/8	Clay loam	WE, FI, SB	HA	FR	SST and SPL	Few F	FC
Bw3	60–90	–	10 YR 5/8	Clay loam	WE, FI, SB	SHA	FR	ST and PL	Few F	FC
Profile P4: Inceptisols, humic dystrudepts										
Ap	0–25	c s	7.5 YR 3/3	Clay	WE, FI, SB	HA	FR	ST and PL	Common F	FF
Bw1	25–44	c s	10 YR 4/4	Clay	WE, FI, SB	HA	FR	ST and PL	Common F	–
Bw2	44–97	g s	7.5 YR 4/6	Clay	WE, FI, SB	HA	FR	ST and PL	Common F	–
2BC1	97–135	c s	7.5 YR 4/6	Clay loam	WE, FI, SB	HA	FR	ST and PL	Common F	–
2BC2	135–144	–	7.5 YR 4/6	Clay loam	WE, FI, SB	HA	FR	ST and PL	Common F	–
Profile P6: Inceptisols, humic dystrudepts										
A	0–20	c s	10 YR 2/2	Sandy clay loam	WE, FI, SB	HA	FR	ST and PL	Common F	FC, MC
Bw1	20–50	g s	10 YR 5/6	Sandy clay loam	WE, FI, SB	HA	FR	ST and PL	Common F	FF, FM
Bw2	50–90	g s	10 YR 5/8	Clay loam	WE, FI, SB	HA	FR	ST and PL	Common F	FF, FM
Bw3	90–117	g s	10 YR 5/6	Clay loam	WE, FI, SB	HA	FR	ST and PL	Common F	FF, FM
BC	117–150	–	10 YR 5/8	Loam	WE, FI, SB	HA	FR	ST and PL	Common F	FF, FM
Profile P8: Inceptisols, typic dystrudepts										
A	0–24	c s	10 YR 3/2	Clay loam	WE, FI, SB	SO	FR	SST and PL	Common F and M	FC, CC

(continued)

Table 2 (continued)

Horizon	Depth (cm)	Boundary[1]	Color hue, value and chroma (Moist)	Texture	Structure[2]	Consistency[3]			Porosity[4]	Roots[5]
Bw1	24–37	c s	10 YR 4/4	Clay loam	WE, FI, SB	SHA	FR	ST and SPL	Common F	FM, MM
Bw2	37–67	a s	10 YR 4/6	Clay loam	WE, FI, SB	SHA	FR	ST and PL	Common F and M	FF, FM
Bw3	67–130	–	10 YR 5/6	Clay loam	WE, FI, SB	SHA	FR	VST and PL	Common F and M	VFF
Profile P12: Entisols, typic udorthents										
A1	0–7	g s	10 YR 3/4	Loamy sand	WE, FI, SB	LO	FR	SST and SPL	Few F	FF
A2	7–41	g s	10 YR 4/6	Loamy sand	WE, FI, SB	LO	FR	SST and SPL	Few F	FF
A3	41–68	g s	10 YR 5/6	Loamy sand ara>	WE, FI, SB	LO	FR	SST and SPL	Few F	FC
C1	68–101	g s	10 YR 4/6	Sand	WE, FI, SB	LO	FR	SST and SPL	Few F	–
C2	101–125	–	10 YR 3/6	Sandy loam	WE, FI, SB	LO	FR	SST and SPL	Few F	–

[1] **D**—Distinctness, a—abrupt, c—clear, g—gradual, d—diffuse; **T**—Topography, s—smooth, w—wavy, i—irregular, b—broken

[2] **G**—Grade, WE—weak, MO—moderate, ST—strong; **S**—size, FI—fine, ME—medium, CO—coarse; **T**—type, SG—single grain, MA—massive, CR—crumbly, CO—columnar, PR—prismatic, PL—platy, AB—angular blocky, SB—subangular blocky

[3] LO—loose, SO—soft, SHA—slightly hard, HA—hard, VHA—very hard, VFR—very friable, FR—friable, FI—firm, VFI—very firm, SST—slightly sticky, ST—sticky, VST—very sticky, SPL—slightly plastic, PL—plastic, VPL—very plastic

[4,5] **S**—Size, V—very fine, F—fine, M—medium, C—coarse; **Q**—quantity, V— very few, F—few, C—common, M—many

Soil structure along depth in all profiles, i.e., the natural organisation of soil particles, was based on grade, size and type. Structure was weak (WE) with aggregates barely observable, fine (FI) with 5–10 mm dimensions, while type was subangular blocky (SB) with aggregate surfaces both rounded and plane, and corners mostly rounded in all profiles.

3.2 Descriptive Statistics of Soil Parameters

Particle size distribution and their ratios were very variable among soil profiles (Table 3). Mean values were 44.7% for sand, 23.5% for silt, 31.4% for clay, 2.75 for sand/silt, 0.98 for silt/clay and 0.52 for clay ratio. Mean BD was 1.00 Mg m^{-3}; soils were acidic (4.70) and non-saline (EC 0.023 dS m^{-1}). Mean SOC was 1.07% (range 0.32–2.67%) and SOC stock 26.1 Mg C ha^{-1} (range 4.9–60.1 Mg C ha^{-1}). Total N was 0.031% as a mean value, available N 31.3 mg kg^{-1}, available P 15.4 mg kg^{-1} and available K 94.1 mg kg^{-1}. CEC was 13.7 $Cmol_+$ kg^{-1}as mean value, Ca^{2+} 0.94 $Cmol_+$ kg^{-1}, Mg^{2+} 1.51 $Cmol_+$ kg^{-1}, Na^+ 0.48 $Cmol_+$ kg^{-1} and K^+ 0.48 $Cmol_+$ kg^{-1}. Mean base saturation was 30.8% (Table 3).

3.3 Physical Soil Parameters

Distribution with depth of sand, silt, clay showed marked differences in the profiles studied. In P12, sand was always dominant (Fig. 3a) and silt and clay showed the lowest proportions (Fig. 3b, c). Conversely, P4 had the least amount of sand (Fig. 3a), and the highest proportion of clay (Fig. 3c) down to 100 cm. The depth distribution of the different soil particles in soil profiles is in agreement with soil texture described in Table 2.

The sand/silt ratio can be used as an index of soil weathering [28, 29]. Thus, the high sand/silt ratio in profile P12 indicates that weathering is low, and the soil is less developed, in agreement with the classification of this profile (Entisol, see Table 2). Conversely, the lower sand/silt ratios of the other profiles indicate that these soils, and in particular P4, are more developed (Inceptisols, see Table 2). Similarly, the silt/clay ratio is considered an index of the degree of weathering of parent materials, and ratios > 0.15 indicate that parent materials are young, and the degree of weathering is low. Silt/clay ratios in the soils studied ranged between 0.41 and 3.87 for all horizons, indicating that in all profiles the weathering of the parent material is at an initial stage. Clay ratio (Fig. 3f) confirms that P4 and P12 have the greatest and least clay contents, respectively.

Bulk density (Fig. 3g) increased with depth in all profiles, except in the deepest horizons of P4 and P12, probably due to an increase in silt content in both profiles and clay content in P12, in agreement with the changes in soil texture (Table 2).

Table 3 Descriptive statistics of soil properties from soils under shifting cultivation in Mokukchung district, Nagaland, North-east India (n = 23). Soil data (mean ± standard deviation) refer to the entire profile depth

Variable	Mean ± SD	Minimum	Maximum
Sand (%)	44.7 ± 19.4	12.9	92.5
Silt (%)	23.5 ± 8.3	6.0	44.0
Clay (%)	31.4 ± 13.4	1.6	57.2
Sand/Silt	2.75 ± 3.21	0.43	15.41
Silt/Clay	0.98 ± 0.80	0.41	3.87
Clay ratio	0.52 ± 0.33	0.02	1.33
Bulk density (Mg m^{-3})	1.00 ± 0.12	0.81	1.23
pH	4.70 ± 0.32	4.19	5.42
EC (dS m^{-1})	0.023 ± 0.011	0.008	0.045
SOC (%)	1.07 ± 0.65	0.32	2.67
SOC stock (Mg C ha^{-1})	26.1 ± 14.2	4.9	60.1
Total N (%)	0.031 ± 0.012	0.009	0.059
Av. N (mg kg^{-1})	31.3 ± 12.1	9.0	59.0
Av. P (mg kg^{-1})	15.4 ± 10.9	4.5	55.5
Av. K (mg kg^{-1})	94.1 ± 105.3	14.9	480
CEC ($Cmol_+$ kg^{-1})	13.7 ± 5.95	3.33	24.0
Ca^{2+} ($Cmol_+$ kg^{-1})	0.94 ± 0.31	0.57	1.65
Mg^{2+} ($Cmol_+$ kg^{-1})	1.51 ± 0.92	0.36	3.46
Na^+ ($Cmol_+$ kg^{-1})	0.48 ± 0.40	0.11	1.02
K^+ ($Cmol_+$ kg^{-1})	0.48 ± 0.55	0.05	1.84
Base saturation (%)	30.8 ± 18.2	12.0	77.9

3.4 Chemical Soil Parameters

Soil organic carbon (%) was least in P12 and greatest in P4 (except for SOC below 45 cm) and presented a pattern opposite to that of sand content (Fig. 4a). The high SOC in P4 is related to the soil order (P4 is an Inceptisol, while P12 is an Entisol) and to the presence of more developed soil horizons (see Sect. 3.1). Similarly, SOC stock (Mg C ha^{-1}) was lowest in P12 and highest in P4 (Fig. 4b), and the distribution with depth was similar to that of CEC. In addition, SOC content and SOC stocks were in agreement with the color hue (Table 2) that indicates that profile P4 (hue 7.5YR, color dark brown to strong brown) is less yellow and darker than profile 12 (hue 10.5YR, color dark yellowish brown). The surface layer contained greater SOC stocks in P4, while P8 contained more SOC in the deeper parts of the soil profile. Available N was generally greater in the surface horizons and the greatest values were found in P8 (Fig. 4c). Available P was greatest in the surface horizon of P12

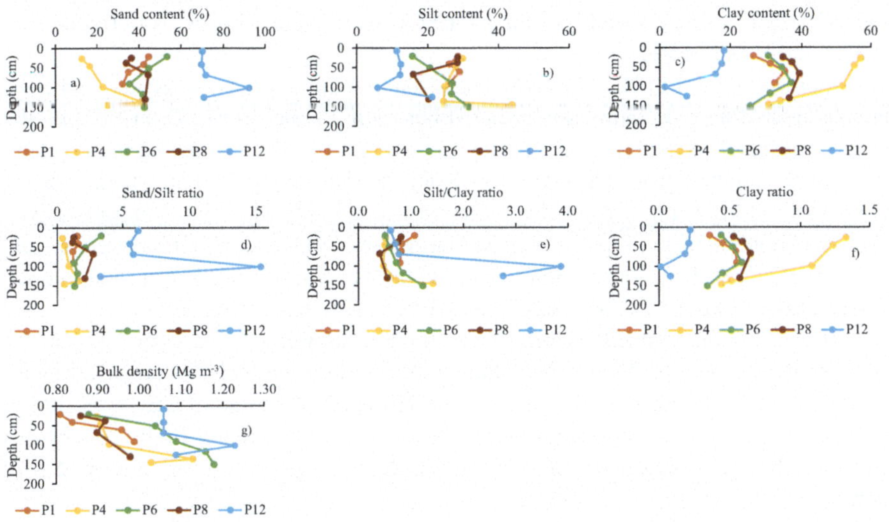

Fig. 3 Depth distribution of physical soil properties in five soil profiles under shifting cultivation in Mokokchung, district, Nagaland, North-east India

(Fig. 4d). Available K generally decreased with depth, showing the greatest values in the topsoil of P8 and P12 and a similar pattern in the other profiles (Fig. 4e).

Compared to the subsoil horizons, A-Ap horizons contained around 37% (33.9 Mg C ha^{-1}) of the SOC stocks, while N, P and K were 1.6, 2.2 and 3.1 times greater,

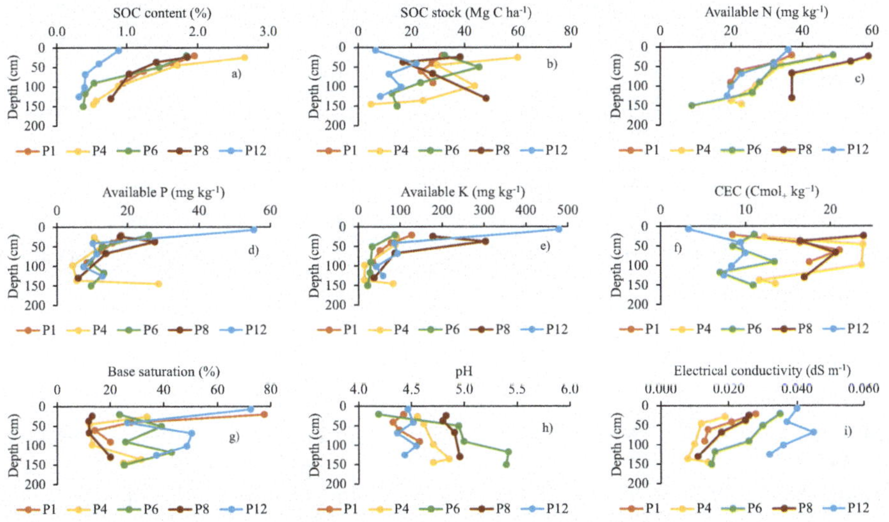

Fig. 4 Depth distribution of chemical soil properties in five soil profiles under shifting cultivation in Mokochung district, Nagaland, North-east India

respectively (Table 4). Notably, values for SOC were greater for the A horizon of P4 (58%) and lower in P12 (11%).

Other findings from Nagaland (Mon and Zunheboto districts) indicated that SOC stocks in shifting cultivation soils ranged between 25 and 30 Mg C ha^{-1} while the range was 30–33 Mg C ha^{-1} under natural forest [30]. Similar results in Nagaland were also obtained from Dimapur and Kohima districts, with SOC stock ranging from 33 to 37 Mg C ha^{-1} in shifting cultivation and 38–48 Mg C ha^{-1} in forests [31]. Indeed, Choudhury et al. [32] reported that in the North-eastern Himalayan region, SOC stocks in shifting cultivation (27 Mg C ha^{-1}) were just 9% lower than in open forest sites (30 Mg C ha^{-1}) across 73 and 148 study sites, respectively.

Cation exchange capacity (Cmol$_+$ kg^{-1}) did not show any defined trend with depth (Fig. 4f), was lower in P6 and P12, and greater in the other profiles consistent with sand content, particularly in the case of P12. However, the trend of CEC was similar to that of SOC stocks. Base saturation (%) was markedly greater in the A horizons of P1 and P12 (Fig. 4g), and similarly to CEC did not show any particular trend with depth.

Soil reaction (pH) decreased with depth in most profiles (Fig. 4h) and P6 was the most acidic in the A horizon (pH 4.19) and the least acidic in the deepest horizons (pH 5.40). Electrical conductivity (EC) decreased with depth (Fig. 4i) except in one

Table 4 Comparison between SOC stocks, N, P and K concentrations in topsoil (A/Ap) and subsoil in five soil profiles under shifting cultivation in Mokochung, district, Nagaland, North-east India, as well as their ratios (A-Ap/subsoil)

	SOC stock (Mg C ha^{-1})		N (mg kg^{-1})		P (mg kg^{-1})		K (mg kg^{-1})	
	A-Ap	Subsoil	A-Ap	Subsoil	A-Ap	Subsoil	A-Ap	Subsoil
P1	31.8	79.9	37.0	24.7	18.4	12.4	128	55.4
A-Ap/subsoil	40		1.5		1.5		2.3	
P4	60.1	103	45.0	26.0	10.8	13.0	92.4	50.9
A-Ap/subsoil	58		1.7		0.8		1.8	
P6	32.6	96.7	49.0	23.8	26.0	11.4	88.6	28.5
A-Ap/subsoil	34		2.1		2.3		3.1	
P8	38.4	93.2	59.0	42.7	18.1	15.9	178	143.4
A-Ap/subsoil	41		1.4		1.1		1.2	
P12	6.6	58.3	36.0	23.5	55.5	10.6	480	71.0
A-Ap/subsoil	11		1.5		5.2		6.8	
Mean values								
Mg C ha^{-1} mg kg^{-1}	33.9	86.2	45.2	28.1	25.7	12.7	194	69.8
A-Ap/subsoil	37		1.6		2.2		3.1	

A-Ap/subsoil value for SOC is expressed in %

horizon of P12. The highest EC was in P12, ranging from 0.032 to 0.045 dS m^{-1}, the lowest in P4 (0.008–0.019 dS m^{-1}).

3.5 *Correlations Between Soil and Topographical Parameters*

Elevation was significantly positively correlated with clay, SOC stock, total N, available N and CEC, negatively with sand and BS. Sand showed a significant negative correlation with silt, clay, and CEC. Clay showed positive correlations with SOC, SOC stocks, total N, available N and CEC, and negative correlations with BD and BS. Bulk density was significantly negatively correlated with SOC, SOC stock, total N, available N, available P and available K. SOC content was positively correlated with SOC stock, total N, available N and CEC, and negatively, but not significantly, with sand content due to the wide variation in sand content (12.9–92.5%). SOC stock was positively correlated with total N, available N and CEC (Table 5).

Results indicated that the SOC stock was greater at higher elevations, positively correlated with clay, SOC content, total N, available N and CEC and negatively with BD. The positive correlation with elevation is a consequence of the lower temperature and the greater soil moisture that slows the rates of SOC decomposition so that more SOC and nitrogen is accumulated [30, 33, 34]. Generally, SOC stocks are greater when clay content is high due to the chemical protection of SOC by clay particles [35, 36]. Positive correlations between SOC and CEC have been reported by other researchers [37, 38], while since BD is greater in sandy soils compared with silty and clayey soils, SOC stocks generally decrease in the order of clay > silt > sand and BD has an inverse relationship with SOC [39, 40].

3.6 *Correlations Between SOC Content and Munsell Color Attributes*

Since SOC has chromogen properties, soil color attributes are a proxy indicator of overall soil quality (the darker the color the greater the SOC, and vice versa) which is related to soil management practices, providing additional information on the soil formation process and other soil characteristics that influence soil color e.g., texture, soil moisture, soil erosion [41]. Results indicated that SOC is significantly negatively correlated with Munsell color attributes (moist soil): value, chroma and their sum (Table 6). Similar results were obtained in from Australia [27], France [27] and Cameroon [28].

Table 5 Spearman's rank correlation matrix of environmental variables and soil properties (n = 23) from soils under shifting cultivation in Mokukchung district, Nagaland, North-east India

Variable	Elevation	Slope	Sand	Silt	Clay	Sand/Silt	Silt/Clay	Clay ratio	BD	pH	EC	SOC	SOC stock	Total N	Av. N	Av. P	Av. K	CEC	Ca^{2+}	Mg^{2+}	Na^{+}	K^{+}
Slope	**0.95**																					
Sand	**−0.61**	**−0.60**																				
Silt	0.41	0.34	**−0.73**																			
Clay	**0.78**	**0.73**	**−0.78**	0.41																		
Sand/Silt	**−0.56**	**−0.53**	**0.96**	**−0.87**	**−0.69**																	
Silt/Clay	**−0.45**	**−0.49**	0.23	0.28	**−0.63**	0.06																
Clay ratio	**0.75**	**0.69**	**−0.81**	0.40	**0.99**	**−0.71**	**−0.63**															
BD	−0.39	−0.24	0.33	−0.27	**−0.48**	0.36	0.37	**−0.48**														
pH	**0.59**	**0.59**	−0.30	0.24	0.39	−0.25	−0.11	0.38	0.36													
EC	**−0.61**	**−0.58**	**0.74**	**−0.49**	**−0.62**	**0.66**	0.18	**−0.64**	0.09	**−0.46**												
SOC	0.35	0.23	−0.36	0.26	**0.54**	−0.36	**−0.46**	**0.53**	**−0.89**	−0.27	−0.05											
SOC stock	**0.49**	0.40	−0.33	0.09	**0.63**	−0.29	**−0.59**	**0.63**	**−0.62**	0.09	−0.32	**0.72**										

(continued)

Table 5 (continued)

Variable	Elevation	Slope	Sand	Silt	Clay	Sand/Silt	Silt/Clay	Clay ratio	BD	pH	EC	SOC	SOC stock	Total N	Av. N	Av. P	Av. K	CEC	Ca^{2+}	Mg^{2+}	Na^+	K^+
Total N	**0.49**	0.40	−0.13	0.02	**0.45**	−0.11	**−0.54**	**0.42**	**−0.74**	−0.05	0.18	**0.77**	**0.53**									
Av. N	**0.47**	0.38	−0.13	0.03	**0.45**	−0.12	**−0.53**	**0.42**	**−0.74**	−0.06	0.18	**0.77**	**0.54**	**1.00**								
Av. P	0.04	−0.03	0.08	0.17	−0.12	−0.01	0.11	−0.14	**−0.44**	−0.29	0.37	0.37	−0.26	**0.50**	**0.49**							
Av. K	−0.02	−0.06	0.13	−0.02	−0.05	0.06	−0.10	−0.06	**−0.55**	**−0.50**	**0.52**	**0.50**	−0.04	**0.66**	**0.67**	**0.72**						
CEC	**0.65**	**0.55**	**−0.69**	0.38	**0.74**	**−0.63**	−0.38	**0.75**	**−0.55**	0.13	**−0.62**	**0.43**	**0.49**	0.27	0.27	−0.12	0.01					
Ca^{2+}	−0.39	−0.35	**0.52**	**−0.64**	−0.40	**0.62**	−0.01	−0.39	0.26	−0.21	0.41	−0.20	−0.38	−0.14	−0.15	0.10	0.21	−0.31				
Mg^{2+}	−0.09	−0.15	**0.47**	−0.09	−0.35	0.33	0.37	−0.37	0.12	0.13	0.39	−0.18	0.04	0.04	0.05	−0.05	−0.03	−0.36	−0.25			
Na^+	−0.16	−0.04	**−0.44**	0.30	0.12	**−0.43**	0.06	0.13	−0.08	−0.28	−0.30	0.13	0.08	−0.30	−0.30	−0.22	−0.16	0.12	−0.15	−0.35		
K^+	−0.10	−0.12	**−0.46**	0.40	0.21	**−0.48**	−0.05	0.23	**−0.57**	**−0.54**	−0.21	**0.59**	0.28	0.10	0.12	0.20	0.21	0.25	−0.18	**−0.44**	**0.72**	
BS	**−0.69**	**−0.60**	**0.58**	−0.29	**−0.67**	**0.52**	0.41	**−0.68**	0.40	−0.26	**0.56**	−0.27	−0.32	−0.28	−0.27	0.02	0.01	**−0.87**	0.22	**0.44**	0.18	0.01

Values in bold are significant at $p < 0.0002$ (Bonferroni correction for multiple comparisons)
Units are the same as in Table 3

Table 6 Spearman's rank correlation matrix between SOC content and Munsell color attributes (n = 23) from soils under shifting cultivation in Mokukchung district, Nagaland, North-east India

Variable	Value	Chroma	Value + Chroma
Chroma	**0.83**		
Value + Chroma	0.92	**0.98**	
SOC (%)	−0.53	**−0.74**	**−0.70**

Values in bold are significant at $p < 0.008$ (Bonferroni correction for multiple comparisons)

4 Conclusions

In the present study, major soil physico-chemical properties were measured through the soil profile in five shifting cultivation areas of Mokukchung, Nagaland. We are not aware of any other studies on the effects of soil type on soil quality and the depth distribution of soil parameters in soils under shifting cultivation. Thus, we specifically aimed to assess the effect of soil type on SOC, the relationship between SOC and some soil parameters, and the correlations among edaphic factors, topography, and soil color. The studied soils belong to the Inceptisol and Entisol orders. SOC content and SOC stock were greatest in an Inceptisol and lowest in an Entisol, which is consistent with the more developed soil horizons in the former. SOC was found to be positively correlated with clay, total N, available N and CEC, while the opposite relationship was found with BD. In addition, SOC had a significant negative correlation with soil color attributes i.e., value and chroma. The present work can serve as baseline study, which will help in monitoring the changes in soil health due to shifting cultivation practices. Moreover, our findings indicate that shifting cultivation in Nagaland might lead to significant changes in soil properties, which will directly or indirectly affect SOC concentrations and stocks. This might suggest avoiding short rotation fallows on Entisols, given their less fertile profile and their reduced capacity to retain nutrients. Moreover, to increase productivity in shifting cultivation, additional sources of nutrients from organic manures and/or adoption of agroforestry with nitrogen-fixing trees can be explored to add nutrients and develop sustainable management of these soils. Detailed studies on the spatial variability of SOC and soil orders for better guiding decision-making on sustainable soil management in shifting cultivation areas of the Eastern Himalayan region are therefore recommended.

References

1. Patel T, Karmakar S, Sanjog J, Kumar S, Chowdhury A (2013) Socio-economic and environmental changes with transition from shifting to settled cultivation in North-Eastern India: an ergonomics [sic] perspective. Int J Agr Sci Res 3:117–136
2. Yadav PK (2013) Slash-and-burn agriculture in North-east India. Expert Opin Environ Biol 2:1. https://doi.org/10.4172/2325-9655.1000102

3. Mishra G, Das PK, Borah R, Dutta A (2017) Investigation of phytosociological parameters and physico-chemical properties of soil in tropical semi-evergreen forests of Eastern Himalaya. J For Res 28:513–520. https://doi.org/10.1007/s11676-016-0330-2
4. Ray SK, Chatterjee D, Rajkhowa DJ, Baishya SK, Hazarika S, Paul S (2020) Effects of integrated farming system and rainwater harvesting on livelihood improvement in North-Eastern region of India compared to traditional shifting cultivation: evidence from an action research. Agrofor Syst 94:451–464. https://doi.org/10.1007/s10457-019-00406-3
5. Jenny H (1994) Factors of soil formation: a system of quantitative pedology. Dover Publications Inc., New York, USA
6. Blanco-Canqui H, Shapiro CA, Wortmann CS, Drijber RA, Mamo M, Shaver TM, Ferguson RB (2013) Soil organic carbon: the value to soil properties. J Soil Water Conserv 68(5):129A-134A. https://doi.org/10.2489/jswc.68.5.129
7. Lal R (2010) Managing soils and ecosystems for mitigating anthropogenic carbon emissions and advancing global food security. Bioscience 60:708–721. https://doi.org/10.1525/bio.2010.60.9.8
8. Wang S, Huang M, Shao X, Mickler RA, Li K, Ji J (2004) Vertical distribution of soil organic carbon in China. Environ Manage 33:S200–S209. https://doi.org/10.1007/s00267-003-9130-5
9. Wiesmeier M, Urbanski L, Hobley E, Lang B, von Lützow M, Marin-Spiotta E, van Wesemael B, Rabot E, Ließ M, Garcia-Franco N, Wollschläger U, Vogel H-J, Kögel-Knabner I (2019) Soil organic carbon storage as a key function of soils—a review of drivers and indicators at various scales. Geoderma 333:149–162. https://doi.org/10.1016/j.geoderma.2018.07.026
10. Batjes NH (1996) Total carbon and nitrogen in the soils of the world. Eur J Soil Sci 47:151–163. https://doi.org/10.1111/j.1365-2389.1996.tb01386.x
11. Jamir A (2014) Shifting options: a case study of shifting cultivation in Mokokchung district in Nagaland, India. In: Erni C (ed) Shifting cultivation, livelihood and food security: new and old challenges for indigenous people in Asia. Food and Agriculture Organization of the United Nations and International Work Group for Indigenous Affairs and Asia Indigenous Peoples Pact, Bangkok, Thailand, pp 159–202
12. Singh AK, Bordoloi LJ, Kumar M, Hazarika S, Parmar B (2014) Land use impact on soil quality in eastern Himalayan region of India. Environ Monit Assess 186:2013–2024. https://doi.org/10.1007/s10661-013-3514-7
13. Soil Survey Staff (2014) Keys to soil taxonomy, 12th edn. United States Department of Agriculture Natural Resources Conservation Service, Washington, DC, USA
14. FAO (2006) Guidelines for soil description, 4th edn. Food and Agriculture Organization of the United Nations, Rome, Italy
15. Gee GW, Bauder JW (1986) Particle-size analysis. In: Klute A (ed) Methods of soil analysis part 1: Physical and mineralogical methods, 2nd edn. American Society of Agronomy, Inc. and Soil Science Society of America, Inc., Madison, WI, USA, pp 383–411. https://doi.org/10.2136/sssabookser5.1.2ed.c15
16. Burt R (ed) (2004) Soil survey laboratory methods manual, soil survey investigations report No. 42. United States Department of Agriculture Natural Resources Conservation Service, Washington, DC, USA
17. Munsell (1975) Standard soil color charts. Macbeth Division of Kollmorgen, Baltimore, MD, USA
18. Blake GR, Hartge KH (1986) Particle-size analysis. In: Klute A (ed) Methods of soil analysis part 1: Physical and mineralogical methods, 2nd edn. American Society of Agronomy, Inc. and Soil Science Society of America, Inc., Madison, WI, USA, pp 363–375. https://doi.org/10.2136/sssabookser5.1.2ed.c15
19. Walkley A, Black IA (1934) An examination of the Degtjareff method for determining soil organic matter and a proposed modification of the chromic acid titration method. Soil Sci 37:29–38
20. Bremner JM (1960) Determination of nitrogen in soil by the Kjeldahl method. J Agric Sci 55:11–33. https://doi.org/10.1017/S0021859600021572

21. Bray RH, Kurtz LT (1945) Determination of total, organic, and available forms of phosphorus in soils. Soil Sci 59:39–46
22. Subbiah B, Asija CL (1956) A rapid procedure for the estimation of available nitrogen in soils. Curr Sci 25:256–260
23. Mishra G, Giri K, Jangir A, Francaviglia R (2020) Projected trends of soil organic carbon stocks in Meghalaya state of Northeast Himalayas, India. Implications for a policy perspective. Sci Total Environ 698:134266. https://doi.org/10.1016/j.scitotenv.2019.134266
24. Jackson ML (1958) Soil chemical analysis. Prentice Hall Inc., Englewood Cliffs, NJ, USA
25. Piper CS (1944) Soil and plant analysis. Interscience Publishers Inc., NY, USA
26. Richards LA (ed) (1954) Diagnosis and improvement of saline and alkali soils, agricultural handbook. United States Department of Agriculture, Washington, DC, USA
27. Viscarra Rossel RA, Minasny B, Roudier P, McBratney AB (2006) Colour space models for soil science. Geoderma 133:320–337. https://doi.org/10.1016/j.geoderma.2005.07.017
28. Kome GK, Enang RK, Yerima BPK (2021) Soil organic carbon distribution in a humid tropical plain of Cameroon: interrelationships with soil properties. Appl Environ Soil Sci 6052513. https://doi.org/10.1155/2021/6052513
29. Van Wambeke AR (1962) Criteria for classifying tropical soils by age. Eur J Soil Sci 13:124–132. https://doi.org/10.1111/j.1365-2389.1962.tb00689.x
30. Mishra G, Francaviglia R (2021) Land uses, altitude and texture effects on soil parameters. A comparative study in two districts of Nagaland, Northeast India. Agriculture 11:171. https://doi.org/10.3390/agriculture11020171
31. Mishra G, Jangir A, Francaviglia R (2019) Modeling soil organic carbon dynamics under shifting cultivation and forests using RothC model. Ecol Model 396:33–41. https://doi.org/10.1016/j.ecolmodel.2019.01.016
32. Choudhury BU, Fiyaz AR, Mohapatra KP, Ngachan S (2016) Impact of land uses, agrophysical variables and altitudinal gradient on soil organic carbon concentration of North-Eastern Himalayan Region of India. Land Degrad Dev 27:1163–1174. https://doi.org/10.1002/ldr.2338
33. Du B, Kang H, Pumpanen J, Zhu P, Yin S, Zou Q, Wang Z, Kong F, Liu C (2014) Soil organic carbon stock and chemical composition along an altitude gradient in the Lushan Mountain, subtropical China. Ecol Res 29:433–439. https://doi.org/10.1007/s11284-014-1135-4
34. Lugo AE, Sanchez MJ, Brown S (1986) Land use and organic carbon content of some subtropical soils. Plant Soil 96:185–196. https://doi.org/10.1007/BF02374763
35. Nichols JD (1984) Relation of organic carbon to soil properties and climate in the southern Great Plains. Soil Sci Soc Am J 48:1382–1384. https://doi.org/10.2136/sssaj1984.03615995004800060037x
36. Six J, Conant RT, Paul EA, Paustian K (2002) Stabilization mechanisms of soil organic matter: implications for C-saturation of soils. Plant Soil 241:155–176. https://doi.org/10.1023/A:1016125726789
37. Khaledian Y, Brevik EC, Pereira P, Cerdà A, Fattah MA, Tazikeh H (2017) Modeling soil cation exchange capacity in multiple countries. Catena 158:194–200. https://doi.org/10.1016/j.catena.2017.07.002
38. Nayak DC, Gangopadhyay SK, Sarka DK, Sen TK (2002) Characteristics and classification of some acid soils of lower Subansiri district of Arunachal Pradesh. Agropedology 12:112–121
39. Prévost M (2004) Predicting soil properties from organic matter content following mechanical site preparation of forest soils. Soil Sci Soc Am J 68:943–949. https://doi.org/10.2136/sssaj2004.9430
40. Mestdagh I, Lootens P, Van Cleemput O, Carlier L (2006) Variation in organic-carbon concentration and bulk density in Flemish grassland soils. J Plant Nutr Soil Sci 169:616–622. https://doi.org/10.1002/jpln.200521861
41. Ibáñez-Asensio S, Marques-Mateu A, Moreno-Ramón H, Balasch S (2013) Statistical relationships between soil colour and soil attributes in semiarid areas. Biosyst Eng 116:120–129. https://doi.org/10.1016/j.biosystemseng.2013.07.013

Recovery of Soil Organic Carbon During Forest Fallow Following Shifting Cultivation in the Eastern Ghats and North-East Hill Region of India

H. C. Hombegowda, M. Madhu, Praveen Jakhar, S. M. Vanitha, and Md. Basit Raza

Abstract Shifting cultivation is conducted by many tribes inhabiting tropical and sub-tropical regions around the world. In India, it is prominently practiced in the Eastern Ghats and North-east hill region states. The major environmental effects of shifting cultivation are loss of forest cover, accelerated soil erosion and land degradation. However, these negative effects are ameliorated by the fallow period which offers the opportunity for recovery of the vegetation and soil fertility. This study aimed to assess the soil organic carbon (SOC) dynamics in the shifting cultivation fallow system and to evaluate the optimum fallow length for SOC recovery. During the study, we quantified the rate at which SOC stocks to 0.5 m depth recovered among fallow periods of increasing age. We measured and compared SOC stocks in natural forest, current shifting cultivation, and seven different aged fallow sites covering a broad range of biophysical conditions across seven locations in three Indian states. Short term cultivation after clearing the natural or secondary forest resulted in sharp declines in the SOC stocks across all locations, with a loss of 55–71% in the Eastern Ghats and 39–41% in the North-east hill region. The recovery of SOC showed a positive relationship with the fallow duration. In the Eastern Ghats, after 50 year fallow periods, SOC stock recovered 60–86% in comparison to the natural forest's SOC. The recovery of SOC was faster in North-east India as SOC recovery after 50 years of fallow was comparable to the natural forest SOC stock. We conclude that

---

H. C. Hombegowda (✉)
ICAR-Indian Institute of Soil and Water Conservation, Research Centre, Sunabeda, Odisha, India
e-mail: hombegowdaars@gmail.com

M. Madhu
ICAR-Indian Institute of Soil and Water Conservation, Dehradun, Uttarakhand, India

P. Jakhar
ICAR-Central Institute for Women in Agriculture, Bhubaneswar, Odisha, India

S. M. Vanitha
ICAR-Indian Institute of Soil and Water Conservation, Research Centre, Udhagamandalam, Tamil Nadu, India

Md. Basit Raza
ICAR-Directorate of Floricultural Research, Pune, Maharashtra, India

S. K. Tripathi and F. Q. Brearley (eds.), *Shifting Cultivation Systems*,
https://doi.org/10.1007/978-3-031-70388-1_7

recovery of SOC stocks showed a positive relationship with the fallow duration but the rate of recovery varied with topographic and climatic factors. During the recovery process, the adjacent natural forests act as seed source and directly contribute to the vegetation establishment and the more rapid recovery of SOC stocks. Among the study sites, SOC stock recovery was significantly faster in the North-eastern hills than the Eastern Ghats.

Keywords Agroforestry · Fallow · Hill agriculture · Soil organic carbon · Tribal agriculture

1 Introduction

'Slash-and-burn' agriculture (generally known as shifting cultivation and locally known as 'jhum' in the North-east hill region and 'podu' in the Eastern Ghats) is a low input subsistence farming system, practiced by communities in the humid and sub-humid tropics [1]. According to recent estimates, 0.26% (0.85 Mha) of the total geographical area of India is under shifting cultivation, predominantly in nine states, viz. Andhra Pradesh, Arunachal Pradesh, Assam, Mizoram, Manipur, Meghalaya, Nagaland, Odisha and Tripura [2]. In remote hilly regions, shifting cultivation is considered as the principal source of livelihoods and regarded as the driving force for the socio-economic status and the tribal economy. Many festivals and other rituals of Indigenous groups revolve around the shifting cultivation fields, because the tribal populations view shifting cultivation not just a means of their livelihood, but as a way of life. Shifting cultivation is generally done on the hill slopes, since land on the foothills belongs to wealthier people and land in valley or plains has an increased chance of crop damage due to flooding [3, 4].

In tropical regions, many of the essential nutrients are released in the form of ash when vegetation is burned. The addition of essential nutrients during the burning phase was developed to overcome the acidic soil and nutrient deficiencies of tropical regions [5]. After the land is cleared, it may be cultivated for a few years without using external inputs. The rapid mineralisation of organic matter and addition of ash after forest clearing and burning provides an increase in available nutrients to the first crop [5]. But the conversion of natural ecosystems especially tropical forests and secondary forests to agricultural ecosystems leads to reductions in soil organic carbon (SOC) by about 25–50% of their initial content [6, 7]. Due to lack of external inputs like organic manure and the high soil erosion rate in the sloping terrain, the fertility of land quickly decreases thus resulting in a reduction of crop yield within two or three years of cultivation. Hence, the farmers have to move to other fields, clearing new secondary forest land for cultivation [8, 9]. The decrease in crop yields, however, is often due to a decline in soil fertility, which may result from a decline in the amount of organic matter in the soil, changes in the physical condition of the soil, and reductions in its nutrient status. In order to be sustainable, shifting cultivation requires large amounts of forested land because the former fields have to be left fallow for several years to restore soil fertility. In recent years, to hasten soil recovery and to improve crop productivity, some farmers are adopting new conservation measures [10]. Previous studies reported that, under moist lowland evergreen or semi-deciduous forest, the greatest amount of soil organic matter is attained after long periods of fallow (e.g. twenty to thirty years) [11]. Information on SOC dynamics during the fallow period after abandoning shifting cultivation fields is of particular interest for initiatives aiming to develop strategies for their management. But most studies are concentrated on vegetation recovery with fewer studies on soil dynamics. Because the results of studies on soil fertility and SOC recovery in a particular agro-climate may not be easily extrapolated to other regions, this can be addressed through an extensive study in each particular region. Hence, in this study, we examined SOC changes during the conversion of forest to shifting cultivation and recovery of SOC during the subsequent fallow as fallow age increased across three Indian states.

2 Materials and Methods

2.1 Study Area

The study was conducted across three states namely Odisha in the Eastern Ghats region and Arunachal Pradesh and Meghalaya in North-east India (Fig. 1; Table 1). Both regions have a (sub)tropical monsoon climate, with a rainy monsoon season from May to November and a distinct dry season from December to April. The mean annual precipitation of the Eastern Ghats ranges from 1265 to 1638 mm and the North-east hill region ranges from 1536 to 12,455 mm while the mean annual temperature (MAT) ranges between 21.9 °C in the highlands to 27.2 °C at lower elevations. In the Eastern Ghats, the soil are moderately acidic (pH around 5.5–6.0) and the predominant soil order is Udic Paleustalfs, whereas in the North-east hill region the soils are more acidic (pH around 4.0–5.0) and commonly belong to the order Inceptisols.

2.2 Sampling Design

In this study, we investigated how SOC stocks changed when forests were cleared for cultivation and current shifting cultivation lands are abandoned and left to fallow with the regrowth of native vegetation allowed. A single plot cluster (location) consisted of nine plots: a natural forest, current shifting cultivation field and seven fallow (secondary forest) sites of different age (5, 10, 20, 30, 40, 50 and > 50 yr). Five plot clusters were established in the central Eastern Ghats and two plot clusters in the North-east hill region (Fig. 1; Table 1). Plot size was 20 m × 20 m (400 m^2) for

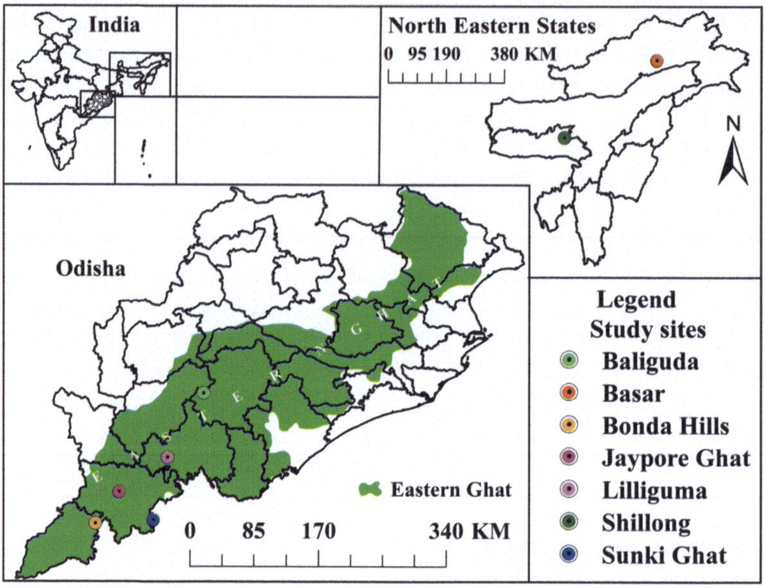

Fig. 1 Location of the study clusters for the examination of changes in soil organic carbon following shifting cultivation in the Eastern Ghats and North-east states hill region of India

observation of vegetation and soil sample collection. In total 70 plots were established (7 plot clusters × 10 treatments). Firstly, the clusters were established randomly around the undisturbed patch of natural forests and in most of the clusters, forest plots were established in the community-managed forests, locally called sacred groves. These are natural forests located near or within the vicinity of the village and they are managed by the local communities for their environmental services and have religious importance. To avoid edge effects, the forest plots were selected at least 40 m from the forest edge. The other plots within the cluster were chosen to only include plots close (< 1 km) to the reference plots with similar biophysical conditions (e.g. soil properties, soil colour, elevation, slope, aspect etc.) and other landscape properties. Accordingly, using a space-for-time substitution sampling approach, any changes observed in SOC may be directly attributable to land-use changes. Only sites which had deep (> 1 m) and well-drained soils were chosen. In each plot, we enumerated all trees ($\geq$ 10 cm diameter) and calculated their relative abundance as their proportional contribution to the total number of tree stems.

Table 1 Elevation and relative abundance of important tree species in abandoned shifting cultivation sites in the Eastern Ghats and North-east hill region of India

Region	Cluster	Elevation (m)	Important tree species and their relative abundance
Eastern Ghats	Jaypore Ghat	870	*Xylia xylocarpa* (Fabaceae) (11.1), *Holarhena pubescens* (Apocynaceae) (6.0), *Santalum album* (Santalaceae) (6.0), *Shorea robusta* (Dipterocarpaceae) (4.5) and *Syzygium cumini* (Myrtaceae) (4.5)
	Sunki Ghat	630	*Xylia xylocarpa* (12.5), *Moringa concanensis* (Moringaceae) (5.8), *Strychnos* sp. (Loganiaceae) (4.7), *Tecoma stans* (Bignoniaceae) (4.8) and *Zanthoxylum asiaticum* (Rutaceae) (4.4)
	Bonda Hills	940	*Xylia xylocarpa* (8.8), *Macaranga peltata* (Euphorbiaceae) (6.1), *Lagerstroemia parviflora* (Lythraceae) (6.1), *Syzygium cumini* (5.6) and *Mangifera indica* (Anacardiceae) (4.4)
	Lilliguma	610	*Shorea robusta* (25.9), *Macaranga peltata* (9.4), *Pterocarpus marsupium* (Fabaceae) (7.7), *Holigarna arnottiana* (Anacardiaceae) (5.1) and *Aglaia elaeagnoidea* (Meliaceae) (3.7)
	Baliguda	540	*Shorea robusta* (11.4), *Xylia xylocarpa* (8.6), *Diospyros truncata* (Ebenaceae) (8.3), *Terminalia bellirica* (Combretaceae) (5.1) and *Ziziphus xylopyrus* (Rhamnaceae) (4.3)
North-east hill region	Shillong	960	*Palaquium polyanthum* (Sapotaceae) (12.3), *Kayea floribunda* (Calophyllaceae) (6.8), *Syzygium cumini* (5.8), *Artocarpus chama* (4.8) and *Zizphus rugosa* (4.4)
	Basar	710	*Dipterocarpus turbinatus* (Dipterocarpaceae) (10.8), *Syzygium cumini* (8.2), *Shorea robusta* (5.2), *Schleichera oleosa* (Sapindaceae) (4.4) and *Neolitsea cassia* (Lauraceae) (4.1)

2.3 Soil Sampling and Analyses

In each 20 m × 20 m plot, we took replicated soil samples using a soil auger from 12 fixed locations around the plot at four soil depths (0–10, 10–20, 20–30 and 30–50 cm). The replicated samples from each depth were pooled and thoroughly mixed. A soil pit (1 × 1 × 1 m) was dug in the centre of the plot where soil cores were taken for soil bulk density determination at 5, 10, 25 and 40 cm depth. Total soil carbon and nitrogen concentrations were analysed using a C/N Elemental Analyser (LECO TruSpec Micro CHN analyser, LECO Corporation, Michigan, USA). The soil carbon stock (Mg C ha^{-1}) was calculated by multiplying the carbon concentration (g kg^{-1}) by the bulk densities of the respective depth interval (kg m^{-3}) and the layer thickness (m) and multiplied by a factor of 10,000 to upscale to one hectare [12].

Total soil carbon stocks for the top 50 cm of soil were calculated as the sum of all depth intervals. To ensure comparability of plot pairs, and to avoid overestimation of SOC stock changes, we used the bulk density data of the respective forest plots to calculate the soil carbon stock of the shifting cultivation sites within each cluster [13, 14]. Furthermore, we recorded information on slope, aspect and elevation of each plot.

2.4 Statistical Analysis

To estimate the size of SOC stock changes following conversion of forest to shifting cultivation and from shifting cultivation to fallow, we calculated the relative difference in SOC stocks between a given plot and the corresponding forest reference of the plots' cluster. The percent difference in SOC stock between forest and other land-use systems was then expressed as the relative change to the respective reference (i.e. natural forest) SOC stock (Relative change = $(SOC_{converted} - SOC_{reference})/SOC_{reference} \times 100$).

3 Results and Discussion

3.1 Forest Conversion for Cropping Reduces SOC Stocks

Our chronosequence study revealed that shifting cultivation after clearing the forest resulted in a sharp decline in the SOC stocks across the seven study clusters, leading to a loss of 55–71% of the original forest SOC stock in the Eastern Ghats and 39–41% in the North-eastern hill region (Figs. 2 and 3). Greater loss of SOC due to shifting cultivation in the Eastern Ghats than the North-east hill states may be due to the high soil erosion rates derived from the greater rainfall intensity and steeper slopes in the studied locations of the Eastern Ghats. In addition, Eastern Ghats farmers practice monocropping that causes more runoff [15], whereas the North-east farmers adopt mixed cropping and in-situ resource conservation practices like the use of wood or remaining trees as in-field barriers that helps to reduce the rate of soil erosion which, in turn, reduces the loss of SOC during the cultivation phase. Additionally, cultural practices such as zero tillage, seed dibbling and the land preparation method adopted in the North-east hill region minimises run off and soil loss, as well SOC loss during cultivation. In a similar chronosequence study, Hombegowda et al. [6] reported that conversion of forest to agriculture in southern India caused 50–61% loss of SOC in the top metre of soil. In a meta-analysis, Powers et al. [16] reported a loss of up to 76% of SOC across global tropical regions due to forest conversion for agriculture use, with losses depending on the precipitation pattern and clay minerology of the soil.

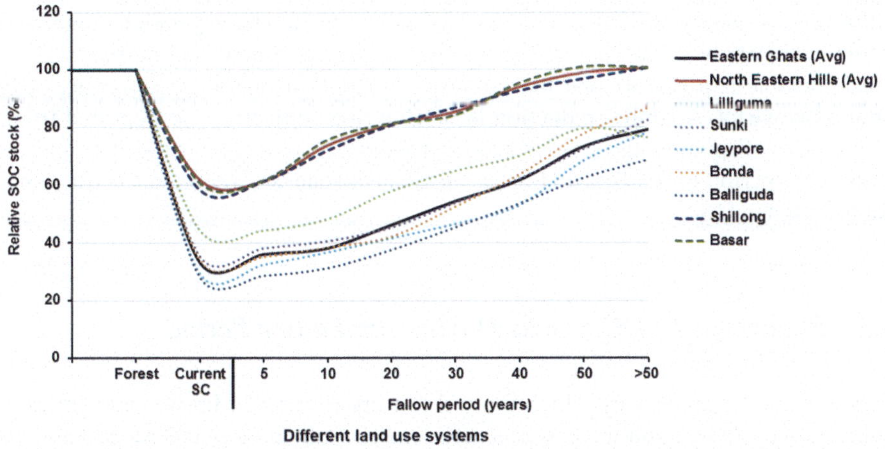

Fig. 2 Relative soil organic carbon (SOC) stock change in the chronosequence from forest—current shifting cultivation (SC)—fallow period with increasing ages in the Eastern Ghats and North-east hill region of India including the average (Avg.) for the two regions

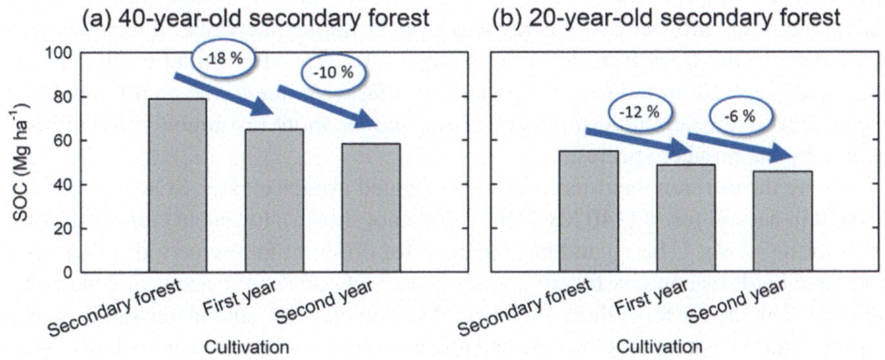

Fig. 3 Change in SOC stocks from secondary forest (40- or 20-years-old) conversion during the first and second year of shifting cultivation in Lilliguma in the Eastern Ghats of India

The decline in SOC during the cropping period was most rapid during the first year but lower in the second year. Our observations in the Lilliguma cluster (Eastern Ghats) revealed a SOC loss of 14.1 and 6.5 Mg C ha^{-1} yr^{-1} when converting a 40-year old fallow to shifting cultivation, during the first and second year of cultivation, respectively, whereas the loss was reduced to 6.4 and 3.0 Mg C ha^{-1} yr^{-1}, respectively, when a 20-year-old fallow was cultivated (Fig. 3). In this example, it appears that the greater the initial SOC stock, the greater the loss in the first year of cultivation. The reduction in SOC during cultivation is mainly due to the faster decomposition process resulting from soil disturbance and accelerated runoff and soil loss from the slopes. The greater amount of soil loss during the first year of cultivation is mainly attributed to the lack of soil protection by vegetation combined with the above given reasons. In response to the rapid decline in soil nutrients, crop yields decline appreciably and the farmer is forced to abandon the cultivated site after a few years. A similar reduction in SOC during cultivation was reported from Mizoram by Wapongnungsang et al. [17] and about 27% reduction in the SOC during a one-year period in the Khasi Hills of Meghalaya state was reported by Toky and Ramakrishnan [18].

3.2 Recovery of SOC Stocks During the Fallow Period

Fallowing is the process by which the soil fertility is restored for the next shifting cultivation cycle by natural forest regeneration. The fallowing of cultivated fields is the most important element in distinguishing shifting cultivation from other types of farming. Nutrients are taken up by the fallow vegetation from the deeper soil horizons and deposited on the soil surface through litterfall. Thus, the increase in SOC is the net increase between the annual litterfall addition and the losses by microbial consumption and leaching [19]. In shifting cultivation sites, grasses, weeds and fast-growing pioneer tree species start to grow immediately after the cultivation, in addition to the trees from left over stumps [20, 21]. After initial establishment of pioneer vegetation, the recovery occurs at a fast rate due to the continuous seed dispersal from the neighbouring forest areas, and the more favourable microclimate created by the pioneer species.

Among the different land-use types investigated in each cluster, SOC stocks were greatest in natural forest (140 Mg C ha^{-1}; data not shown), lowest in current shifting cultivation (33 Mg C ha^{-1}) and intermediate for different fallow periods of shifting cultivated lands (secondary forests). Among the SOC stocks of current shifting cultivation sites in the Eastern Ghats, the lowest SOC stock was found in the Sunki cluster (33 Mg C ha^{-1}), whereas at the current shifting cultivation sites in Bonda Hills, SOC was greatest and nearly half (45%) compared with the associated reference natural forest (Fig. 4). In the North-east hill regions, current shifting cultivation sites had SOC nearly double (74–81 Mg C ha^{-1}) that of the Eastern Ghats sites (Fig. 4). The recovery of SOC stocks during the fallow period was dependent on the duration of fallow—showing an increase as the fallow period lengthened. But there was a high variation for SOC recovery among each fallow period age between clusters (Fig. 4) that can be attributed to differences in topography, soil nutrient status and vegetation composition, among others.

The different secondary vegetation establishment on fallowed shifting cultivated lands in all seven locations may have caused variation in SOC stock accumulation (Figs. 2 and 4). Up to 30 years, the SOC accumulation rate was slower (50–68 Mg C ha^{-1}) and, in all clusters studied, the greatest SOC stock gain was recorded between 30 and 50 years of fallow age. After 50 years of fallow in the North-eastern states, the SOC stock reached a value close to the forest values (98.9 $\pm$ 2.9%). But during

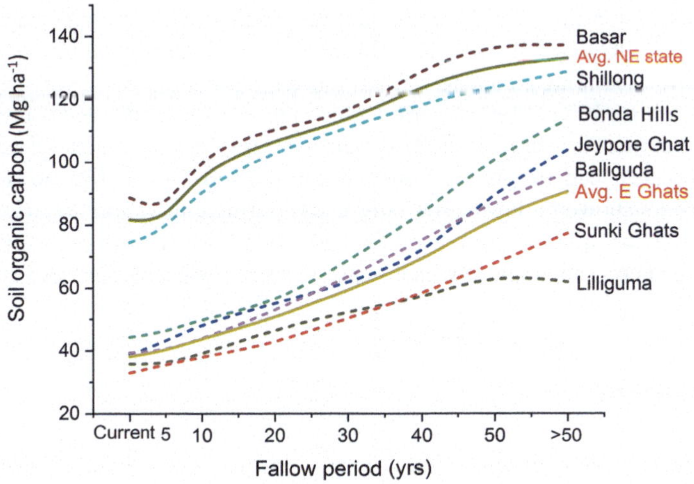

Fig. 4 Soil organic carbon stock (Mg C ha^{-1}) recovery under increasing fallow period age in the Eastern Ghats and North-east hill region states of India including the average (Avg.) for the two regions

the same fallow period in the Eastern Ghats, SOC stock reached only 78.7 ± 6.8% of the forest values.

A chronosequence study of 36 fallow sites carried out in the tropical dry forest of the southern Yucatan Peninsula region in Mexico concluded that total tree basal area increased with fallow age and achieved its maximum during old fallow ages [22]. Similar results were reported by Peña-Claros [23] from tropical moist forest fallows in Bolivia. A meta-analysis by Poorter et al. [24] covering about 1500 secondary forest plots in the Neotropical region found that secondary forests recover biomass slowly in the initial years but can reach up to 122 Mg C ha^{-1} during 20 years of fallow. These studies showed that the vegetation regrows slowly during the initial years and reaches nearly 50% of the natural forest biomass during long fallow periods [25, 26]. The lower basal area of the vegetation during the initial years provides less litter input but a greater density of larger trees, and increased species richness in old fallow sites are likely to harbour species with strong resource-utilisation characteristics compared with less species-intensive systems [27]. This may promote greater net primary production [28], which, in turn, promotes soil C sequestration. Over longer fallow periods, a more complete utilisation of resources takes place and the system becomes more productive [29]. These theories support the positive relationship between above ground biomass and SOC stocks in the old fallow sites. The relationship between vegetation and SOC stock was reported by Hombegowda et al. [6] where they found a direct relationship between tree basal area and SOC stocks in southern Indian forests. As previous researchers found, our study also found that SOC stock gain was slower during the initial fallow years (up to 30 years) and increased to a maximum during the 30–50 year fallow periods (Figs. 2 and 4). This indicates that intensity of litter production and root activity are important aspects in relation to SOC gain [30] which are slow to increase during the initial period of fallow recovery. Among the Eastern Ghats clusters, Bonda Hills showed the greatest SOC stock after a 50-year fallow period which may be due to the seed source (forest area) surrounding the study area. This helps in rapid establishment of vegetation which ultimately leads to greater litter input to the soil. In general, the North-east Indian clusters showed a lower SOC stock loss during cultivation due to the lower rainfall intensities in the region, use of felled wood logs as barriers to soil erosion and the mixed cropping pattern which conserves soil resources more effectively.

4 Conclusions

Long fallow periods with greater tree density and less steep slopes contribute to a faster SOC recovery in former shifting cultivation sites of India. The greatest SOC stock was found in the 50-year fallow indicating that a considerable period of time is needed for substantial SOC recovery. Shorter fallow periods (< 10 years) reduced the recovery potential of the land, resulting in a slow recovery of vegetation and SOC. In steeply sloping areas, shifting cultivation leads to degradation of land and replacement of multi-storey vegetation with pure stands of dominant tree species like *Shorea robusta* (Dipterocarpaceae). In steeply sloping areas of the Eastern Ghats, the SOC recovery is incomplete even after 50 years of fallow.

References

1. Cairns M, Garrity DP (1999) Improving shifting cultivation in Southeast Asia by building on indigenous fallow management strategies. Agrofor Syst 47:37–48. https://doi.org/10.1023/A:1006248104991
2. Wastelands Atlas of India (2019) National Remote Sensing Centre, Indian Space Research Organisation, Department of Space, Government of India Hyderabad, India
3. Jha LK (1997) Shifting cultivation. APH Publishing Corporation, New Delhi, India
4. Pandey DK, De HK, Dubey SK, Kumar B, Dobhal S, Adhiguru P (2020) Indigenous people's attachment to shifting cultivation in the Eastern Himalayas, India: a cross-sectional evidence. For Policy Econ 111:102046. https://doi.org/10.1016/j.forpol.2019.102046
5. Hombegowda HC, Jakhar P, Madhu M, Marwein Y (2021) 'Thang bun': indigenous practice of *in situ* biochar preparation-cum-application for improved jhum cultivation in North East India. Curr Sci 120:1160–1168. https://doi.org/10.18520/cs/v120/i7/1160-1168
6. Hombegowda HC, van Straaten O, Köhler M, Hölscher D (2016) On the rebound: soil organic carbon stocks can bounce back to near forest levels when agroforests replace agriculture in southern India. Soil 2:13–23. https://doi.org/10.5194/soil-2-13-2016
7. Lal R (2005) Soil carbon sequestration in natural and managed tropical forest ecosystems. J Sustain For 21:1–30. https://doi.org/10.1300/J091v21n01_01
8. Jordan CF (1987) Shifting cultivation. Case study No 1: Slash and burn agriculture near San Carlos de Rio Negro, Venezuela. In: Jordan CF (ed) Amazonian rain forests: ecosystem disturbance and recovery. Springer Verlag, New York, USA, pp 9–23

9. Roder W, Phengchanh S, Maniphone S (1997) Dynamics of soil and vegetation during crop and fallow period in slash-and-burn fields of northern Laos. Geoderma 76:131–144. https://doi.org/10.1016/S0016-7061(96)00100-0
10. Hombegowda HC, Adhikary PP, Jakhar P, Madhu M, Barman D (2020) Hedge row inter cropping impact on run-off, soil erosion, carbon sequestration and millet yield. Nutr Cycl Agroecosyst 116:103–116. https://doi.org/10.1007/s10705-019-10031-2
11. Grogan P, Lalnunmawia F, Tripathi SK (2012) Shifting cultivation in steeply sloped regions: a review of management options and research priorities for Mizoram state, Northeast India. Agrofor Syst 84:163–177. https://doi.org/10.1007/s10457-011-9469-1
12. Veldkamp E (1994) Organic carbon turnover in three tropical soils under pasture after deforestation. Soil Sci Soc Am J 58:175–180. https://doi.org/10.2136/sssaj1994.03615995005800010025x
13. de Koning GHJ, Veldkamp E, López-Ulloa M (2003) Quantification of carbon sequestration in soils following pasture to forest conversion in northwestern Ecuador. Glob Biogeochem Cycl 17:1098. https://doi.org/10.1029/2003GB002099
14. de Blécourt M, Brumme R, Xu J, Corre MD, Veldkamp E (2013) Soil carbon stocks decrease following conversion of secondary forests to rubber (*Hevea brasiliensis*) plantations. PLoS ONE 8:e69357. https://doi.org/10.1371/journal.pone.0069357
15. Lenka NK, Dass A, Sudhishri S, Patnaik US (2012) Soil carbon sequestration and erosion control potential of hedgerows and grass filter strips in sloping agricultural lands of eastern India. Agric Ecosyst Environ 158:31–40. https://doi.org/10.1016/j.agee.2012.05.017
16. Powers JS, Corre MD, Twine TE, Veldkamp E (2011) Geographic bias of field observations of soil carbon stocks with tropical land-use changes precludes spatial extrapolation. Proc Natl Acad Sci USA 108:6318–6322. https://doi.org/10.1073/pnas.1016774108201l
17. Wapongnungsang, Manpoong C, Tripathi SK (2018) Changes in soil fertility and rice productivity in three consecutive years cropping under different fallow phases following shifting cultivation. Int J Plant Soil Sci 25:46087. https://doi.org/10.9734/IJPSS/2018/46087
18. Toky OP, Ramakrishnan PS (1981) Cropping and yields in agricultural systems of the northeastern hill region of India. Agro-Ecosyst 7:11–25. https://doi.org/10.1016/0304-3746(81)90012-3
19. Lemma B, Berggren Kleja D, Olsson M, Nilsson I (2007) Factors controlling soil organic carbon sequestration under exotic tree plantations: a case study using the CO2Fix model in southwestern Ethiopia. For Ecol Manage 252:124–131. https://doi.org/10.1016/j.foreco.2007.06.029
20. Neba NE (2009) Cropping systems and post-cultivation vegetation successions: agroecosystems in Ndop, Cameroon. J Hum Ecol 27:27–33. https://doi.org/10.1080/09709274.2009.11906188
21. Styger E, Rakotondramasy HM, Pfeffer MJ, Fernandes ECM, Bates DM (2007) Influence of slash-and-burn farming practices on fallow succession and land degradation in the rainforest region of Madagascar. Agri. Ecosys Environ 119:257–269. https://doi.org/10.1016/j.agee.2006.07.012
22. Read L, Lawrence D (2003) Recovery of biomass following shifting cultivation in dry tropical forests of the Yucatan. Ecol Appl 13:85–97. https://doi.org/10.1890/1051-0761(2003)013[0085:ROBFSC]2.0.CO;2
23. Peña-Claros M (2003) Changes in forest structure and species composition during secondary forest succession in the Bolivian Amazon. Biotropica 35:450–461. https://doi.org/10.1646/01078
24. Poorter L, Bongers F, Aide TM, Almeyda Zambrano AM, Balvanera P, Becknell JM, Boukili V, Brancalion PHS, Broadbent EN, Chazdon RL, Craven D, de Almeida-Cortez JS, Cabral GAL, de Jong BHJ, Denslow JS, Dent DH, DeWalt SJ, Dupuy JM, Durán SM, Espírito-Santo MM, Fandino MC, César RG, Hall JS, Hernández-Stefanoni JL, Jakovac CC, Junqueira AB, Kennard D, Letcher SG, Licona J-C, Lohbeck M, Marín-Spiotta E, Martínez-Ramos M, Massoca P, Meave JA, Mesquita R, Mora F, Muñoz R, Muscarella R, Nunes YRF, Ochoa-Gaona S, de Oliveira AA, Orihuela-Belmonte E, Peña-Claros M, Pérez-García EA, Piotto D, Powers

JS, Rodríguez-Velázquez J, Romero-Pérez IE, Ruíz J, Saldarriaga JG, Sanchez-Azofeifa A, Schwartz NB, Steininger MK, Swenson NG, Toledo M, Uriarte M, van Breugel M, van der Wal H, Veloso MDM, Vester HFM, Vicentini A, Vieira ICG, Vizcarra Bentos T, Williamson GB, Rozendaal DMA (2016) Biomass resilience of Neotropical secondary forests. Nature 530:211–214. https://doi.org/10.1038/nature16512
25. Ruiz J, Fandiño MC, Chazdon RL (2005) Vegetation structure, composition, and species richness across a 56-year chronosequence of dry tropical forest on Providencia Island, Colombia. Biotropica 37:520–530. https://doi.org/10.1111/j.1744-7429.2005.00070.x
26. Lalnunzira C, Tripathi SK (2018) Leaf and root production, decomposition and carbon and nitrogen fluxes during stand development in tropical moist forests, north-east India. Soil Res 56:306–317. https://doi.org/10.1071/SR16265
27. Tilman D, Lehman CL, Thomson KT (1997) Plant diversity and ecosystem productivity: theoretical considerations. Proc Natl Acad Sci USA 94:1857–1861. https://doi.org/10.1073/pnas.94.5.1857
28. Vandermeer J (1989) The ecology of intercropping. Cambridge University Press, Cambridge, UK
29. Kirby KR, Potvin C (2007) Variation in carbon storage among tree species: implications for the management of a small-scale carbon sink project. For Ecol Manage 246:208–221. https://doi.org/10.1016/j.foreco.2007.03.072
30. Jackson RB, Schenk HJ, Jobbágy EG, Canadell J, Colello GD, Dickinson RE, Field CB, Friedlingstein P, Heimann M, Hibbard K, Kicklighter DW, Kleidon A, Neilson RP, Parton WJ, Sala OE, Sykes MT (2000) Belowground consequences of vegetation change and their treatment in models. Ecol Appl 10:470–483. https://doi.org/10.1890/1051-0761(2000)010[0470:BCOVCA]2.0.CO;2

Differences in Soil Carbon and Nitrogen Across Contrasting Land-Use Systems in a Sub-Tropical Hilly Region, North-East India

Etsoshan Y. Ovung, Francis Q. Brearley, Ng. Somen Singh, and Shri Kant Tripathi

Abstract This study examines the impact of land-use change on soil properties in a hilly sub-tropical region in North-east India. We collected soil samples from three depths from five dominant land-uses, namely, natural forests (NAF), current jhum (shifting cultivation) (CJ), jhum fallows (JF), homegardens (HG) and *Acacia pennata* plantations (AP) that were analysed for their carbon (C) and nitrogen (N) concentration and stocks. Our results indicated that soil properties were significantly influenced by both land-use and soil depth. The greatest C and N concentrations were found in NAF compared to JF, CJ, AP, and particularly HG, indicating that the least disturbed soils had the best soil quality. Notably, however, C and N stocks were not less under shifting cultivation (i.e. CJ or JF) compared to NAF, although they were less under AP and HG. This study reflected distinct differences in soil properties under land-use change, particularly more intensive land-uses such as plantations, ultimately leading to negative feedbacks between soil properties and land-use systems.

Keywords Land-use change · Soil health · Soil organic matter

1 Introduction

Anthropogenic pressures are leading to land degradation and loss of ecosystem services. A key ecosystem service provided by soil is that of carbon (C) storage, with C stocks in Indian soils being around 63 Pg (to a depth of 1.5 m) [1] and with mountainous regions of India having proportionally greater stocks of C due to the greater above-ground vegetation cover and reduced soil temperatures slowing

E. Y. Ovung · Ng. S. Singh · S. K. Tripathi
Department of Forestry, School of Earth Science and Natural Resources Management, Mizoram University, Aizawl 796004, India

F. Q. Brearley (✉)
Department of Natural Sciences, Manchester Metropolitan University, Chester Street, Manchester M1 5GD, UK
e-mail: f.q.brearley@mmu.ac.uk

S. K. Tripathi and F. Q. Brearley (eds.), *Shifting Cultivation Systems*,
https://doi.org/10.1007/978-3-031-70388-1_8

decomposition [1, 2]. Additionally, stocks of soil nitrogen (N) are often linked to C due to them both being majorly derived from organic matter, and are fundamental to maintaining vegetation primary production as N is the major limiting nutrient for the majority of ecosystems around the world [3].

In Mizoram, one of the states in North-east India, landscapes are characterised by a high forest cover but the primary source of livelihood for the majority of the rural population is shifting cultivation (jhum) [4] which, with long fallow periods, is a sustainable practice through which the regrowth of secondary forests regenerates the soil C and nutrient status. Soil health is clearly important for successful cultivation, but in Mizoram, which is dominated by steep slopes and torrential rains during the rainy monsoon season, soil erosion through water and other forms of land degradation is estimated to affect 34.9% of the total geographical area of the state (21,081 km^2) [5].

There has been a rapid change in land-use in North-east India. Complementing the traditional shifting cultivation fields and fallows, and homegardens, there has been more recent expansion of horticultural areas and plantations which are well-known to influence soil C stocks, particularly through disturbance [6–10]. Appropriate soil management practices and other integrated land-use management activities are thus important in maintaining soil physical, chemical and biological soil properties. It is now crucial to evaluate the impacts of contrasting land-use activities on soil properties and fertility for sustainable agricultural productivity. We therefore initiated this study with the aim to quantify the patterns of soil C and N across five prominent land-use systems of Mizoram, and through the soil profile. The five land-uses studied included natural forests, current jhum fields, jhum fallows, homegardens, and *Acacia pennata* (Fabacaee) plantations.

2 Materials and Methods

2.1 Study Area

Mizoram is a hilly state with a sub-tropical to moist climate (Köppen's Cwa) experiencing heavy monsoonic rainfall from May to September that totals up to 3000 mm per annum. The most common soil order found in the state are Inceptisols that are generally composed of sandy loams and clayey loams having high C content [11]. The study sites, namely natural forest (NAF), current jhum fields (CJ), jhum fallows of at least 12 years of age (JF), homegardens (HG), and *Acacia pennata* (AP) plantations, were found within the two adjoining districts of Aizawl and Mamit. The precise geographical location and more general description of the study sites, including vegetation composition, is provided in Ovung [12].

2.2 Experimental Design and Sampling

In each of the five land-use systems, three replicated sites measuring about 50 m × 50 m with similar topography, soil and vegetation were established, totaling 15 study sites [13]. Soil samples were collected randomly from each of the sites by inserting a 10 cm soil corer with an inner diameter of 5.2 cm. Within each of the sites, ten soil cores were collected and samples were then drawn from three depths, i.e. 0–10 cm, 10–20 cm and 20–30 cm and composited into two samples per depth for further analysis. Thus, a total of 90 samples (5 land-use types × 3 replicates × 3 depths × 2 composite samples) were collected. The soil samples were air-dried and roots were removed before soils were ground, and passed through a 2-mm sieve prior to further analysis.

2.3 Analysis of Soil Physical and Chemical Properties

Soil bulk density was estimated from undisturbed soil using a core sampler having a known volume (noted above), after oven drying the samples at 105 °C till constant mass. Soil colour of fresh samples was determined using a Munsell soil colour chart [14]. Dried and sieved samples (c. 0.25 g) were used for the determination of total C and N using a Leco TruSpec CN elemental analyser.

2.4 Calculations and Statistical Analyses

Soil C and N stocks (Mg ha^{-1}) were calculated with following formula:

$$\text{Stock}\left(\text{Mg}\,\text{ha}^{-1}\right) = 10^4\left(\text{m}^2\,\text{ha}^{-1}\right) \times \text{soil depth (m)} \times \text{bulk density}\left(\text{Mg}\,\text{m}^{-3}\right) \times \text{concentration}\ (\%)/100.$$

Data obtained were subjected to two-ways analysis of variance (ANOVA), one-way ANOVA with Tukey's test, or Spearman's correlation coefficients using Minitab v.19.2, as appropriate.

3 Results

Soil colour ranged from brown to dark yellowish brown in the surface layer, while the deeper layers were dominated by light yellowish brown to brownish yellow colours (Table 1). Bulk density increased with depth ($F = 24.9$, $p < 0.001$) and was significantly less dense in the NAF compared to all the other land-use types ($F = 27.4$, $p < 0.001$) (Table 1).

Soil C concentrations were significantly greater in the NAF compared to all other land-use types, and least in HG (Fig. 1). Soil C declined significantly with increasing soil depth but the magnitude of this decline was greater in the less-disturbed land-use types (i.e., NAF, JF and CJ). Total N followed a similar pattern to C, but had very similar concentrations in the two deeper soil layers under AP. Differences in the C:N ratio were less marked between land-uses and were greater in NAF and JF; there was no significant interaction between land-use and soil depth. Soil C was correlated with both soil colour value ($r = -0.75$, $p < 0.001$) and hue ($r = 0.48$, $p = 0.001$).

Table 1 Soil colour and bulk density (mean ± standard error) at three different soil depths across different land-use systems in Mizoram, North-east India (NAF = Natural forest, CJ = Current jhum, JF = Jhum fallow, HG = Homegarden, and AP = *Acacia pennata* plantation)

Land-use (depth in cm)	Munsell colour notation			Soil colour	Bulk density (g cm^{-3})
	Hue	Value	Chroma		
NAF					
(0–10)	10 YR	4	3	Brown	0.92 ± 0.03
(10–20)	10 YR	6	8	Brownish yellow	1.05 ± 0.05
(20–30)	10 YR	7	8	Yellow	1.15 ± 0.05
CJ					
(0–10)	7.5 YR	5	3	Brown	1.17 ± 0.02
(10–20)	10 YR	6	8	Light yellowish brown	1.24 ± 0.03
(20–30)	10 YR	6	6	Light yellowish brown	1.31 ± 0.06
JF					
(0–10)	10 YR	4	4	Dark yellowish brown	1.13 ± 0.02
(10–20)	10 YR	6	6	Brownish yellow	1.22 ± 0.03
(20–30)	10 YR	7	6	Yellow	1.25 ± 0.02
HG					
(0–10)	10 YR	4	6	Dark yellowish brown	1.24 ± 0.02
(10–20)	10 YR	6	4	Light yellowish brown	1.31 ± 0.02
(20–30)	10 YR	6	6	Brownish yellow	1.37 ± 0.05
AP					
(0–10)	10 YR	5	3	Brown	1.17 ± 0.03
(10–20)	10 YR	5	6	Yellowish brown	1.27 ± 0.04
(20–30)	10 YR	6	6	Brownish yellow	1.32 ± 0.02

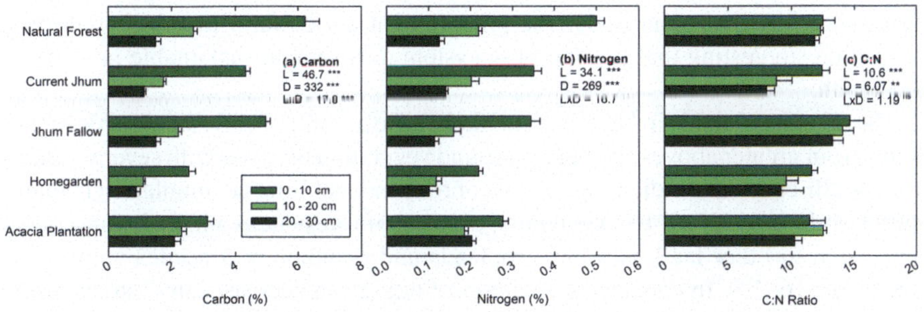

Fig. 1 Soil organic carbon and nitrogen concentrations and carbon:nitrogen ratio (mean ± standard error) at three different soil depths across different land-use systems in Mizoram, North-east India. Results of two-way ANOVAs reported as F-values and *** for $p < 0.001$, ** for $P < 0.01$ and ns for non-significant (L = land-use system, D = soil depth, L × D = land-use system × soil depth)

Table 2 Soil organic carbon and nitrogen stocks (Mg ha^{-1}; mean ± standard error) at two different soil depths (0–10 cm and 0–30 cm) across different land-use systems in Mizoram, North-east India (NAF = Natural forest, CJ = Current jhum, JF = Jhum fallow, HG = Homegarden, and AP = *Acacia pennata*)

	Depth	NAF	CJ	JF	HG	AP
Carbon	0–10 cm	57.5 ± 4.56 a	50.9 ± 1.58 a	56.0 ± 2.14 a	31.6 ± 2.29 b	36.7 ± 3.22 b
	0–30 cm	103 ± 5.83 a	87.0 ± 2.41 a	101 ± 4.30 a	58.0 ± 4.24 b	93.6 ± 3.12 a
Nitrogen	0–10 cm	4.63 ± 0.31 a	4.13 ± 0.26 ab	3.87 ± 0.29 ab	2.73 ± 0.09 c	3.23 ± 0.07 bc
	0–30 cm	8.38 ± 0.32 a	8.46 ± 0.43 a	7.25 ± 0.67 ab	5.62 ± 0.45 b	8.26 ± 0.15 a

Letters in a row indicate significant differences according to a Tukey's test with $p < 0.05$

Total C stocks in the top 10 cm of soil were significantly less in HG and AP compared to NAF, whilst C stocks in the top 30 cm were only less in HG when compared to NAF (Table 2). The same patterns were seen in the case of N stocks (Table 2).

4 Discussion

Soil C and N play a major role in the productivity of soil and are both dependent on land management practices and the amount and quality of organic matter input from the land-use system in question. The results from the present study revealed significant negative effects on soil C and N when comparing native forest to various other land-use systems. Notably, however, the effects were greater when land-uses were more intensive than under the predominant agricultural practice of shifting cultivation suggesting that this land-use system may remain sustainable in terms of soil health.

The greatest C and N in NAF is most likely to be a result of elevated organic matter inputs from greater above- and below-ground vegetation biomass with several studies from North-east India indicating the role of root and leaf litter accumulation in maintaining soil C and N [15, 16]. Generally, clearing of forest areas and their subsequent conversion to other land-uses like croplands and plantations concedes a considerable C loss [6–10]. In concurrence with our findings on decreases in C stocks when comparing native forest to other land-uses, van Straaten et al. [17] also reflected that conversion of natural forest to land-uses such as rubber (*Hevea brasiliensis*; Euphorbiaceae) and oil palm (*Elaeis guineensis*; Arecaceae) plantations lead to decreases in C stocks by 50%. Veldkamp et al. [8] reviewed deforestation impacts on soil C stocks in the tropics reporting an average loss of 17% of C in the top 10 cm of cropland soils compared to native forests during the initial 10 years after deforestation. This C loss can be attributed to more rapid organic matter decomposition as a result of alterations to factors such as temperature, soil water content and aeration. In contrast, less C in HG can be related to continuous cultivation and the sparse tree cover along with the consistent removal of the surface vegetation through weeding and clearing of fallen leaf litter. Sahoo et al. [18] found soil C concentrations of 2.4% in the topsoil of a range of large (c. 0.8 ha) Mizoram homegardens but the concentration in smaller (c 0.1 ha) homegardens was about half this (1.3%) possibly due, again, to greater organic matter input in larger homegardens with more and larger trees. Another important factor contributing to the loss of N in cultivated soils such as HG can be related to the uptake of N by crops since N is one of the most in-demand nutrients for crop growth [3].

It is notable that differences in total C stocks were more pronounced in the top 10 cm of soil compared to the top 30 cm where only HG showed less C compared to NAF, this is a common pattern in studies of land-use change effects on soils [8, 9]. This pattern may also be due to the greater relative accumulation of C in the upper horizons in the less disturbed land-uses such as NAF and JF in this study. These patterns can also be determined by examining the soil colour which is a reasonably good indicator of soil C concentration as exemplified by Ferrando Jorge et al. [19]. These authors indicated that soil colour lightness was a more effective predictor of soil C, when compared to chroma or hue, with a linear decrease in soil lightness as SOC increased.

Reforestation by means of secondary succession can lead to reversal of C loss [8] and enhance the C:N ratio [8] as in the case of JF in this study. In some cases, plantations may also lead to C accumulation [20, 21] but this is more likely if they are planted on degraded lands. Further examination of plantation species-specific effects would be valuable in follow-on research. We found similar N concentrations in the two deeper soil layers of the *Acacia pennata* plantations that may be linked to the rhizospheric nitrogen fixation attributable to the leguminous nature of the planted trees that are able to enhance soil fertility [22]. In other geographically-related studies, Lungmuana et al. [23] found that teak (*Tectona grandis*; Lamiaceae) plantations had a more negative effect on soil C than other plantations. Similarly, Manpoong et al. [24] showed differences in soil C concentrations across various land-use systems with rubber and teak plantations exhibiting lower soil C in comparison to jhum fallows, *Tephrosia candida* (Fabaceae) plantations and natural forests.

5 Conclusion

Increased human activities have significantly influenced soil characteristics and subsequently the sustainability of agro-ecosystems. Clearing of native forests for cultivation negatively impacts soil properties as indicated by this study. Important soil fertility indicators such as SOC and total N were significantly affected by many of the land-use systems studied with homegardens having the least fertile soils. In contrast, we see that jhum fallow of at least 12 years of age showed soil properties similar to the natural forest as a result of succession. Owing to the tremendous importance of SOC and its influence on soil health, management for its retention and accumulation, e.g. the '4 per mille' initiative [25], is a key factor in developing management policies for contrasting land-use systems in the region.

Acknowledgements We thank the Ministry of Tribal Affairs, Government of India, for providing financial assistance in the form of a National Fellowship for Higher Education of Scheduled Tribe Students. We also express our gratitude to Manchester Metropolitan University for providing funding through a QR GCRF allocation and all necessary resources throughout the course of this work.

References

1. Bhattacharyya T, Pal DK, Mandal C, Velayutham M (2000) Organic carbon stock in Indian soils and their geographical distribution. Curr Sci 79:655–660
2. Choudhury BU, Mohapatra KP, Das A, Das PT, Nongkhlaw L, Fiyaz AR, Ngachan SV, Hazarika S, Rajkhowa DJ, Munda GC (2013) Spatial variability in distribution of organic carbon stocks in the soils of North East India. Curr Sci 104:604–614
3. LeBauer DS, Treseder KK (2008) Nitrogen limitation of net primary productivity in terrestrial ecosystems is globally distributed. Ecology 89:371–379. https://doi.org/10.1890/06-2057.1
4. Grogan P, Lalnunmawia F, Tripathi SK (2012) Shifting cultivation in steeply sloped regions: a review of management options and research priorities for Mizoram state, Northeast India. Agrofor Syst 84:163–177. https://doi.org/10.1007/s10457-011-9469-1
5. Choudhury BU, Nengzouzam G, Ansari MA, Islam A (2022) Causes and consequences of soil erosion in northeastern Himalaya, India. Curr Sci 122:772–789. https://doi.org/10.18520/cs/v122/i7/772-789
6. Don A, Schumacher J, Freibauer A (2011) Impact of tropical land- use change on soil organic carbon stocks—a meta-analysis. Glob Change Biol 17:1658–1670. https://doi.org/10.1111/j.1365-2486.2010.02336.x
7. Powers JS, Corre MD, Twine TE, Veldkamp E (2011) Geographic bias of field observations of soil carbon stocks with tropical land-use changes precludes spatial extrapolation. Proc Natl Acad Sci USA 108:6318–6322. https://doi.org/10.1073/pnas.1016774108

8. Veldkamp E, Schmidt M, Powers JS, Corre MD (2020) Deforestation and reforestation impacts on soils in the tropics. Nat Rev Earth Environ 1:590–605. https://doi.org/10.1038/s43017-020-0091-5
9. Padbhushan R, Kumar U, Sharma S, Rana DS, Kumar R, Kohli A, Kumari P, Parmar B, Kaviraj M, Sinha AK, Annapurna K, Gupta VVSR (2022) Impact of land-use changes on soil properties and carbon pools in India: a meta-analysis. Front Env Sci 9:794866. https://doi.org/10.3389/fenvs.2021.794866
10. Peng Y, Yuan C, Hed nec P, Yue K, Zhu G, Jin Z, Wei S, Wu F (2022) Effects of transforming multiple ecosystem types to plantations on soil carbon, nitrogen, and phosphorus concentrations at the global scale. Plant Soil 481:213–227. https://doi.org/10.1007/s11104-022-05632-w
11. Colney L, Nautiyal BP (2013) Characterization and evaluation of soils of Aizawl district, Mizoram, India using remote sensing and GIS techniques. J Geomat 7:83–91
12. Ovung EY (2021) Changes in soil aggregation, microbial population, carbon and nutrients under different land use systems of Mizoram. PhD Thesis, Mizoram University, India
13. Ovung EY, Tripathi SK, Brearley FQ (2021) Changes in soil exchangeable nutrients across different land uses in steep slopes of Mizoram, North-east India. J Appl Nat Sci 13:929–936. https://doi.org/10.31018/jans.v13i3.2795
14. Munsell Color Co., Inc. (1992) Munsell soil color charts. Baltimore, MD, USA
15. Wapongnungsang, Hauchhum R, Tripathi SK (2017) Litter decomposition *vis-a-vis* carbon and nitrogen dynamics of *Tephrosia candida* components in different fallow periods following shifting cultivation in Mizoram. Indian J Ecol 44:791–796
16. Lalnunzira C, Tripathi SK (2018) Leaf and root production, decomposition and fluxes of carbon and nitrogen during stand development in tropical moist forests, north-east India. Soil Res 56:306–317. https://doi.org/10.1071/SR16265
17. van Straaten O, Corre MD, Wolf K, Tchienkoua M, Cuellar E, Matthews RB, Veldkamp E (2015) Conversion of lowland tropical forests to tree cash crop plantations loses up to one-half of stored soil organic carbon. Proc Natl Acad Sci USA 112:9956–9960. https://doi.org/10.1073/pnas.1504628112
18. Sahoo U, Rocky P, Vanlalhriatpuia K, Upadhyaya K (2010) Structural diversity and functional dynamism of traditional home-gardens of North-east India. Bioscan Spec Issue 1:159–171
19. Ferrando Jorge N, Clark J, Cárdenas ML, Geoghegan H, Shannon V (2021) Measuring soil colour to estimate soil organic carbon using a large-scale citizen science-based approach. Sustainability 13:11029. https://doi.org/10.3390/su7043528
20. de Koning GHJ, Veldkamp E, López- Ulloa M (2003) Quantification of carbon sequestration in soils following pasture to forest conversion in north western Ecuador. Glob Biogeochem Cycles 17:1098. https://doi.org/10.1029/2003GB002099
21. Silver WL, Ostertag R, Lugo AE (2000) The potential for carbon sequestration through reforestation of abandoned tropical agricultural and pasture lands. Restor Ecol 8:394–407. https://doi.org/10.1046/j.1526-100x.2000.80054.x
22. Brockwell J, Searle SD, Jeavons AC, Waayers M (2005) Nitrogen fixation in acacias: an untapped resource for sustainable plantations, farm forestry and land reclamation, ACIAR monograph 115. Australian Centre for International Agricultural Research, Canberra, Australia
23. Lungmuana SSB, Choudhury BU, Vanthawmliana SS, Hnamte V (2019) Transforming jhum to plantations: effect on soil microbiological and biochemical properties in the foot hills of North Eastern Himalayas, India. Catena 177:84–91. https://doi.org/10.1016/j.catena.2019.02.008
24. Manpoong C, Wapongnungang, Tripathi SK (2021) Soil carbon stock in different land-use systems in the hilly terrain of Mizoram, Northeast India. J Appl Nat Sci 13:723–728. https://doi.org/10.31018/jans.v13i2.2615
25. Minasny B, Malone BP, McBratney AB, Angers DA, Arrouays D, Chambers A, Chaplot V, Chen Z-S, Cheng K, Das BS, Field DJ, Gimona A, Hedley CB, Hong SY, Mandal B, Marchant BP, Martin M, McConkey BG, Mulder VL, O'Rourke S, Richer-de-Forges AC, Odeh I, Padarian J, Paustian K, Pan G, Poggio L, Savin I, Stolbovoy V, Stockmann U, Sulaeman Y, Tsui C-C, Vågen T-G, van Wesemael B, Winowiecki L (2017) Soil carbon 4 per mille. Geoderma 292:59–86. https://doi.org/10.1016/j.geoderma.2017.01.002

Ethnopedological Knowledge of Jhumias in Garo Hills, Meghalaya, to Assess Soil Nutrients and Health

Gaurav Mishra and Avishek Sarkar

Abstract Ethnopedology is the study of documenting information from indigenous/local communities towards soil, its classification, usage, and management of fertility. This information on how a particular soil type can be better utilized for particular crops has been developed by farmers through extensive experience, but it is rarely reported. Keeping the importance of ethnopedology in mind, the present study was conducted in the West Garo Hills of Meghalaya where shifting cultivation farmers (jhumias) were interviewed to assess their ethnopedological knowledge for managing soil resources which was then validated through determination of soil fertility indices. Interaction with farmers revealed that *amang* is the common vernacular name of soil in the Garo community and they classify soil into four distinct categories. To assess and validate the soil properties of different soil categories, soil samples (0–30 cm depth) were collected and analysed to calculate the soil fertility index (SFI), soil evaluation factor (SEF) and soil organic carbon (SOC) stocks. The soil analyses showed that bulk density and SOC content and stocks were significantly different among soil categories and when compared with forest soils. However, other soil parameters like soil pH, exchangeable cations and available phosphorus did not differ. The soil health indices, SFI and SEF, were also not significantly different among soil categories although *amang gisim*, after forest soil, is more fertile in terms of SFI, SEF and SOC stock. That might be one of the reasons that jhumias in the Garo Hills prefer to cultivate this soil category with cereals. The ethnopedological knowledge developed by jhumias of the Garo community in Meghalaya is well validated with the help of standard soil analyses and can be incorporated in future soil health assessment studies.

Keywords Ethnopedology · Farmers · Nutrients · Soil health

G. Mishra (✉)
Centre of Excellence on Sustainable Land Management, Indian Council of Forestry Research and Education, Dehradun, Uttarakhand 248006, India
e-mail: gaurav.mishra215@gmail.com

A. Sarkar
ICFRE - Rain Forest Research Institute, Jorhat, Assam 785001, India

S. K. Tripathi and F. Q. Brearley (eds.), *Shifting Cultivation Systems*,
https://doi.org/10.1007/978-3-031-70388-1_9

1 Introduction

In any land-use system, essential plant nutrients remain in continuous and dynamic cycles [1] as they are absorbed from soil by plant roots and used in certain metabolic activities like enzyme activation, synthesis of macromolecules, and regulation of transpiration [2]. Later, nutrients are returned to the soil in the form of litterfall and its subsequent decomposition by a variety of organisms [3]. The amount of nutrients stored in soil will be able to depict the status of soil health at a particular time. The role of soil nutrient availability in controlling the net primary productivity of terrestrial ecosystems is well determined [4]. Nowadays, the concept of soil health is based on the quantitative measurement of physical, chemical or biological properties of soil [5]. Earlier, when quantitative methods of nutrient analysis were not available, local farmers used their indigenous knowledge about soil and its management as land evaluation methods; this was passed on and accumulated through generations [6, 7]. This traditional knowledge about soil, its classification, use, and management is called as ethnopedology [8] and has evolved over millennia [9]. As reported by Gong [10], the earliest known soil classification system was reported in the Chinese book entitled *Yugong* published 2500 years BP.

North-east India is a global biodiversity hot spot [11] and a culturally diverse region of the world [12]. It is a representative example of tropical areas, having steep slopes and receiving high rainfall [13]. Nearly 65% of the total area of North-east India is under forests and the remainder is under crops or remains uncultivated [14]. The local farmers (jhumias) inhabiting the hilly regions of North-east India are dependent on diverse natural resources and successfully manage soils for their livelihood requirements. In hilly states of North-east India, paddy (*Oryza sativa*; Poaceae) is cultivated as staple crop, along with diversity of crops like maize (*Zea mays*; Poaceae), millet (*Eleusine coracana*; Poaceae) and vegetables [15]. Crop diversity in slash-and-burn jhumming systems is high, with a documented variation from 15 to 60 crops in a single field [16]. But nowadays, local communities are more oriented towards settled plantations systems, which they grow as per their experience in a particular soil type. These include modern cash crops such as rubber (*Hevea brasiliensis*; Euphorbiaceae) and tea (*Camellia sinensis*; Theaceae), and traditional cash crops including areca nut (*Areca catechu*; Arecaceae), broom grass (*Thysanolaena latifolia*; Poaceae), and ginger (*Zingiber officinale*; Zingiberaceae) [17, 18]. Indigenous knowledge about soil, its typical characteristics, fertility status on the basis of colour, and how it can be better utilized for a crop is evaluated by jhumias according to their experiences [9, 19]. This traditional knowledge about soil has rarely been reported from North-east India. Therefore, the present study was undertaken in the Garo Hills of Meghalaya, with the following objectives: (1) to collect information about the ethnopedological knowledge of jhumias, (2) to determine how jhumias were managing their soil resources for different crops, and (3) to validate their knowledge with quantitative soil parameters and indices.

2 Materials and Methods

2.1 *Study Area*

The study was conducted in the Rongram block of Garo Hills in Meghalaya state. The Garo Hills are located at 25° 9′–26° 1′ N and 89° 49′–91° 2′ E and occupy an area of about 8170 km^2, with a maximum elevation of 1412 m above mean sea level. The climate of the Garo Hills is mainly controlled by the south-west monsoon and experiences high temperatures for much of the year with an average annual precipitation of 2800–3300 mm [20]. Soils belong to the Alfisols order, Udalfs suborder, Hapludalfs group, UlticHapludalfs subgroup [21].

2.2 *Ethnopedological Data and Soil Sample Collection*

To obtain information about ethnopedological knowledge of jhumias, 50 farmers were informally interviewed whilst working in their fields. During soil sampling, farmers were asked about the different soil categories, their local names, their characteristics, their perception about the soil fertility and overall health, and the kind of management practices they were adopting to cultivate the different soils. After obtaining the information about soil categories and management practices, according to their understanding and traditional knowledge, they were also asked about the preferred crop(s) to be grown in a particular soil category. Thereafter, 50 soil samples from the surface layer (0–30 cm depth) of four different farmer-categorized soil categories (40 samples) were collected in consultation with the farmers along with those from nearby forest (10 samples). This sampling depth (0–30 cm) represents the most dynamic horizon of soil, which is most influenced by cultivation practices. For bulk density (BD) estimation, a metal core was inserted vertically to collect an undisturbed sample, while a 250 g soil sample, from each point was also collected. In the laboratory, soil samples were air-dried, gently ground, and sieved to 2-mm for further chemical analysis.

2.3 *Soil Analysis and Indices Calculation*

Bulk density of soil was calculated after oven drying of samples at 105 °C for 72 h. Soil pH was determined in 1:2.5 soil–water suspension using a digital pH meter (Systronics pH System 361), while the wet oxidation method of Walkley and Black [22] was used to estimate soil organic carbon (SOC) content of soils. Later, soil organic matter (SOM) was calculated using the factor of 1.72, which is commonly used to calculate SOM from SOC. Exchangeable cations [calcium (Ca), potassium (K), and magnesium (Mg)] were extracted with 1 M ammonium acetate (at pH 7.0).

Potassium was determined by a Systronics flame photometer [23], while Ca and Mg were determined by ethylenediaminetetraacetic acid (EDTA) titration given by Schollenberger and Simon [24]. Values for the soil fertility index (SFI) [25] and the soil evaluation factors (SEF) [26] were calculated using the following equations:

$$\begin{aligned} \mathrm{SFI} = {} & \mathrm{pH} + \text{organic matter } (\%) + \text{available P}\left(\mathrm{mg\ kg^{-1}}\right) + \text{exch. K}\left(\mathrm{Cmol_{+}\ kg^{-1}}\right) \\ & + \text{exch. Ca}\left(\mathrm{Cmol_{+}\ kg^{-1}}\right) + \text{exch. Mg}\left(\mathrm{Cmol_{+}\ kg^{-1}}\right) - \text{exch. Al}\left(\mathrm{Cmol_{+}\ kg^{-1}}\right) \end{aligned} \tag{1}$$

$$\begin{aligned} \mathrm{SEF} = {} & [\text{exch. K}\left(\mathrm{Cmol_{+}\ kg^{-1}}\right) + \text{exch. Ca}\left(\mathrm{Cmol_{+}\ kg^{-1}}\right) + \text{exch. Mg}\left(\mathrm{Cmol_{+}\ kg^{-1}}\right) \\ & - \log\left(1 + \text{exch. Al}\left(\mathrm{Cmol_{+}\ kg^{-1}}\right)\right)] \times [\mathrm{SOM}\ (\%) + 5] \end{aligned} \tag{2}$$

SOC stock was calculated with the equation given in Mishra et al. [27]:

$$\text{SOC stock} = \text{SOC } (\%) \times \text{BD} \times \text{D} \tag{3}$$

where SOC indicates the soil organic carbon (%), BD is the bulk density (g cm^{-3}), and D is thickness (cm) of the soil layer in question.

2.4 Statistical Analyses

Soil parameters were summarized with their mean values and standard deviations. To assess the differences in soil parameters among different soil categories, one-way ANOVAs were conducted, followed by the Duncan's test ($P < 0.05$). All the statistical analyses were performed in SPSS version 16.0.

3 Results and Discussion

3.1 Ethnopedological Knowledge and Preferred Crops

Interaction with jhumias of the Garo community reveals that they use the word *amang* for soil, and they have knowledge four distinct categories of soil that they have classified accordingly. Their soil classification was mainly based on colour and appearance (Table 1), and in their vernacular they are as follows: *amang roong* (stony soil), *amang gisim* (black soil), *amang sitapa* (clayey soil), and *amang gitchak* (red soil). As per their knowledge, *roong* is very friable and the least fertile in nature. The notable characteristics of these soils are shallow depth and a predominance of stones with the preferred crops being betel nut (*Areca catechu*), cashew nut (*Anacardium occidentale*; Anacardiaceae) and pineapple (*Ananas comosus*; Bromeliaceae), which

Table 1 Ethnopedological features of four soil categories classified by Garo farmers in Meghalaya, North-east India

Category	Appearance/ colour	Fertility status	Preferred crop	Consistency	Hardness
Roong	Stony	Less fertile	Suitable for plantation crops like betel nut, cashew nut and pineapple	Friable	–
Gisim	Black	Very fertile	All crops	Moderately sticky	Less compact
Sitapa	Clayey	–	Not suitable for cropping	Very sticky	Hard on drying
Gitchak	Red	Cow dung is needed	All crops	Sticky	Less compact

require minimal management except only occasional weeding. The second category of soil is *gisim,* which is very fertile so can be used for growing different cereals and vegetables without any additional source of nutrients. The third one is *sitapa,* which is unsuitable for cropping purposes due to its sticky nature during the rainy season and hardness during dry summers. Lastly, they note *gitchak,* which is mainly red in colour, and can be used for cultivating rubber and tea. Moreover, they mentioned that *gitchak* soils are less fertile in comparison to *gisim* soils and cow dung, as an additional source of nutrients, is used in these soils to obtain good production when cultivated for cereal crops like paddy.

3.2 Soil Parameters

Bulk density (BD) was significantly different among the soil categories with the greatest BD reported in *sitapa* (1.24 g cm^{-3}), followed by *roong* (1.18 g cm^{-3}) and *gitchak* (1.10 g cm^{-3}), and then *gisim* (1.08 g cm^{-3}), while the minimum value was found in forest soils (0.98 g cm^{-3}) (Table 2). Bulk density is known to be affected by the amount of coarse fragments in soil and/or changes in land-use patterns or SOC content [28]. Hajabassi et al. [29] also reported that there was a 20% increase in BD due to cultivation, when 30–40 year old forest had been cropped for 20 years, after deforestation. Our results were also along the same lines, showing lower BD in forest areas with cultivated soil categories like *gisim* and *gitchak* soils having a BD about 10% denser when compared with forest soils. Moreover, lesser BD in forest soils may also be due to their greater SOC content, while the converse can explain the high BD in *sitapa* and *roong* soils. Soils across North-east India are mainly acidic in nature and the pH among different soil categories varied from 4.33

to 5.92. However, there were no statistically significant differences between the pH of different soil categories (Table 2).

In the present study, the SOC content was significantly different among different soil categories (Table 2). The greatest SOC was found in forest soils (2.46%) followed by *gisim* (2.01%), which was statistically similar to *gitchak* (1.77%). In forest ecosystems, litterfall is the main cause of high organic matter and SOC [30]. In contrast, in shifting cultivation fields, after obtaining the economic part of the crops, residues are left in the field to decompose. This acts as a mulch for retaining the moisture as well as increasing the carbon and other nutrients in the soil. The contribution of plant residues and their subsequent decomposition cannot, therefore, be ignored in other cultivated soils, like *gisim*, which is going to influence the SOC content [3]. In *gitchak* soils, mostly rubber and tea were cultivated by the farmers, which can also produce considerable amounts of litter or other carbon inputs [27, 31]. The lowest values of SOC were recorded under *sitapa* (1.25%) and *roong* soils (1.26%), which were statistically similar to each other. The effects of different soil categories on exchangeable cations were not significant (Table 2). The concentrations of Ca, Mg, K and Al varied between 1.65 to 2.09, 1.55 to 1.90, 0.72 to 1.72 and 3.52 to 5.57 $Cmol_+ \, kg^{-1}$, respectively. However, it is important to note that North-east India is characterized by high rainfall with a hilly topography, which can be the reason for the low concentrations of cations found in forest soils, due to leaching of bases from the exchange complexes. Similar to exchangeable cations, the phosphorus concentration did not vary among different soil categories (Table 2).

3.3 Soil Fertility Indices

The soil fertility index (SFI) and soil evaluation factor (SEF) for different soil categories were calculated to validate the ethnopedological knowledge of farmers regarding soil health. Our results derived from soil analyses and indices revealed that, although the mean SFI (18.2) and SEF (20.6) was greatest in *gisim*, followed by forest soil (17.5 and 18.7), values were not statistically different under different soil categories (Fig. 1). According to the farmers' perception, *gisim* is suitable for the majority of the crops they grow, with less requirement for intense management, while *sitapa* soils are uncultivable. Our analytical results are also along the same line, as maximum and minimum values of SFI and SEF were found in *gisim* and *sitapa*, respectively, although our results were not significant, but significant variations in SFI and SEF under different land-uses were reported by other authors [1, 32]. Greater values of SFI and SEF from surface layers of forest soils were also reported by Panwar et al. [1] from West Bengal and Pal et al. [32] from acid soil of the north-western Himalayas. They reported that forest soils are much 'healthier' than other land-uses, due to accumulation of mineral nutrients as a result of decomposition of leaf litter, high SOC content and increased microbial activity.

Table 2 Variation (mean ± standard deviation) in soil chemical properties of four soil categories classified by Garo farmers in Meghalaya, North-east India

Soil category	BD (g cm^{-3})	pH	SOC (%)	Ca ($Cmol_+$ kg^{-1})	Mg ($Cmol_+$ kg^{-1})	K ($Cmol_+$ kg^{-1})	Al ($Cmol_+$ kg^{-1})	P (mg kg^{-1})
Forest soil	0.98 (±0.05)[a]	4.96 (±0.53)	2.46 (±0.27)[c]	1.65 (±0.59)	1.55 (±0.57)	0.72 (±0.52)	4.78 (±2.34)	9.38 (±1.29)
Roong	1.18 (±0.08)[bc]	4.95 (±0.39)	1.26 (±0.17)[a]	1.98 (±0.37)	1.90 (±0.44)	1.58 (±0.56)	5.31 (±2.19)	9.78 (±2.29)
Gisim	1.08 (±0.32)[b]	5.68 (±0.87)	2.01 (±0.61)[b]	2.09 (±0.54)	1.63 (±0.65)	1.72 (±1.57)	4.63 (±3.50)	9.01 (±3.34)
Sitapa	1.24 (±0.07)[c]	5.92 (±0.76)	1.25 (±0.16)[a]	1.90 (±0.63)	1.76 (±0.68)	1.39 (±0.70)	3.52 (±2.32)	10.23 (±3.09)
Gitchak	1.10 (±0.09)[bc]	4.33 (±0.68)	1.77 (±0.16)[b]	1.75 (±0.63)	1.65 (±0.84)	1.33 (±0.99)	5.57 (±2.78)	7.86 (±2.43)
F value	5.48		25.63					
P value	< 0.01		< 0.01					

Values in same column followed by different letters are significantly different at $p < 0.05$ by Duncan's multiple-range test

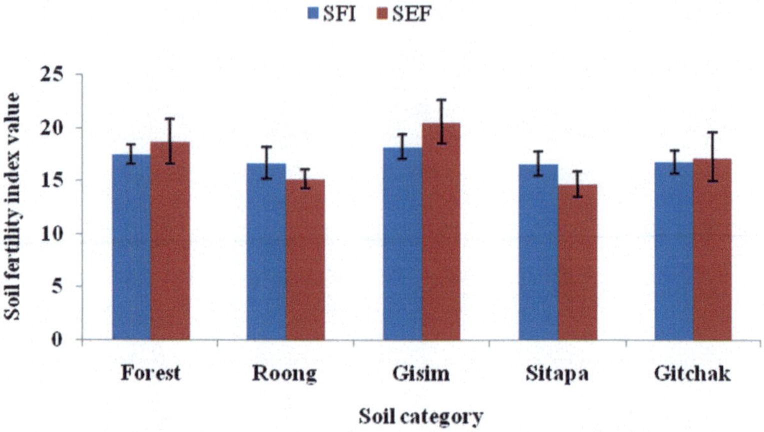

Fig. 1 Mean (± standard deviation) soil fertility index (SFI) and soil evaluation factor (SEF) values under different soil categories as classified by Garo farmers in Meghalaya, North-east India

3.4 *Soil Organic Carbon Stock*

Assessment of SOC stocks is necessary because it is an important ecosystem service which is imperative to understand with respect to land-use change or soil type and serves as a critical soil health indicator. In the present study, SOC stock was assessed to 30 cm depth in different soil categories and compared with forest soils. The differences in SOC stock among different soil categories were found significant: the greatest SOC stock was found in forest soil (69.0 ± 2.67 Mg C ha^{-1}), followed by *gisim* (62.3 ± 4.68 Mg C ha^{-1}) and *gitchak* (57.8 ± 1.39 Mg C ha^{-1}) soils, with the minimum recorded in *sitapa* (46.3 ± 2.04 Mg C ha^{-1}) and *roong* (44.8 ± 2.45 Mg C ha^{-1}) soils. *Gisim* soils are cultivated with various kinds of crops along with legumes which help in N_2 fixation and addition of easily decomposable materials [16]. Inclusion of leguminous cover crops can be recommended for other soil types that can be cultivated. Our results are in line with the findings of Dutta [33], who reported that mixed forests in Meghalaya had greatest SOC storage when compared with grasslands and croplands. Biomass removal and exposure of soils to mineralization and decomposition, coupled with degradation of soil structure [34], were the possible causes of low SOC stocks under cultivated soils (Fig. 2).

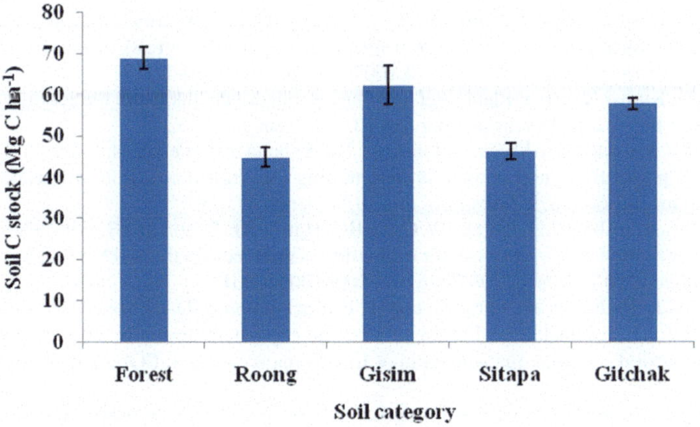

Fig. 2 Mean (± standard deviation) soil carbon stock (t C ha^{-1}) under different soil categories as classified by Garo farmers in Meghalaya, North-east India

4 Conclusions

This present study documents the ethnopedological knowledge developed by local farmers of the Garo community in Meghalaya (North-east India). Their knowledge about soil health showed that the soil they classify as *gisim* is more fertile in comparison to others, and no external inputs were needed to enhance its productivity. However, locally available nutrient management techniques can be used to improve the nutrient status of *gitchak* soil for growing other crops besides plantation crops as currently done. Slightly higher, but not significantly different, values of SFI and SEF and significantly greater SOC stock in forest soils is mainly due to accumulation and decomposition of litter, which maintains the nutrient balance in soil. This study validates that ethnopedological knowledge of farmers through laboratory analyses of soil nutrient status and soil health. Therefore, there is a good rationale to incorporate ethnopedological knowledge in soil quality assessment studies for improved productivity and to guarantee livelihood requirements.

Acknowledgements The authors are thankful to the Indian Council of Forestry Research and Education (ICFRE), Dehradun for supporting this research work through project RFRI/2018-19/SFM-1.

References

1. Panwar P, Pal S, Reza SK, Sharma B (2011) Soil fertility index, soil evaluation factor, and microbial indices under different land uses in acidic soil of humid subtropical India. Commun Soil Sci Plant Anal 42:2724–2737. https://doi.org/10.1080/00103624.2011.622820

2. Mishra G, Das PK, Borah R, Dutta A (2017) Investigation of phytosociological parameters and physico-chemical properties of soil in tropical semi-evergreen forests of Eastern Himalaya. J For Res 28:513–520. https://doi.org/10.1007/s11676-016-0330-2
3. Mishra G, Kumar R (2016) Plant litter decomposition: drivers insight to the ecological process. Eur J Biol Res 6:176–185. https://doi.org/10.5281/zenodo.58752
4. Fernández-Martínez M, Vicca S, Janssens IA, Espelta JM, Peñuelas J (2017) The role of nutrients, productivity and climate in determining tree fruit production in European forests. New Phytol 213:669–679. https://doi.org/10.1111/nph.14193
5. Karlen DL, Mausbach MJ, Doran JW, Cline RG, Harres RF, Schuman GE (1997) Soil quality: a concept definition, and framework for evaluation (a guest editorial). Soil Sci Soc Am J 61:4–10. https://doi.org/10.2136/sssaj1997.03615995006100010001x
6. Dumanski J, Bindraban PS, Pettapiece WW, Bullock P, Jones RJA, Thomasson A (2002) Land classifications, sustainable land management, and ecosystem health. In: Encyclopedia of food and agricultural sciences. Encyclopedia of Life Support Systems, EOLSS Publishers, Oxford, UK
7. Nuwategeka E, Nyeko M (2019) Using indigenous knowledge to model land suitability for crops in Amuru district, northern Uganda. Afr J Food Agric Nutr Dev 19:14805–14828. https://doi.org/10.2136/10.18697/ajfand.87.17555
8. WinklerPrins AMGA, Sandor JA (2003) Local soil knowledge: insights, applications, and challenges. Geoderma 111:165–170. https://doi.org/10.1016/S0016-7061(02)00262-8
9. Nath AJ, Lal R, Das AK (2015) Ethnopedology and soil quality of bamboo (*Bambusa* sp.) based agroforestry system. Sci Total Environ 521–522:372–379. https://doi.org/10.1016/j.scitotenv.2015.03.059
10. Gong Z (1994) Chinese soil taxonomic classification (first proposal). Institute of Soil Science, Academia Sinica, Nanjing, China
11. Myers N, Mittermeier RA, Mittermeier CG, da Fonseca GA, Kent J (2000) Biodiversity hotspots for conservation priorities. Nature 403:853858. https://doi.org/10.1038/35002501
12. Chatterjee S, Saikia A, Dutta P, Ghosh D, Worah S (2006) Review of biodiversity in Northeast India, Background Paper No. 13, WWF-India, Delhi, India
13. Mishra G, Sarkar A (2020) Studying the relationship between total organic carbon and soil carbon pools under different land management systems of Garo hills, Meghalaya. J Environ Manage 257:110002. https://doi.org/10.1016/j.jenvman.2019.110002
14. Saha R, Chaudhary RS, Somasundaram J (2012) Soil health management under hill agroecosystem of North East India. Appl Environ Soil Sci 696174. https://doi.org/10.1155/2012/696174
15. Nagaland Environmental Protection and Economic Development and International Institute of Rural Reconstruction (1999) Building upon traditional agriculture in Nagaland, India. Nagaland Environmental Protection and Economic Development, Nagaland, India and International Institute of Rural Reconstruction, Silang, Cavite, The Philippines
16. Giri K, Mishra G, Rawat M, Pandey S, Bhattacharyya R, Bora N, Rai JPN (2020) Traditional farming systems and agro-biodiversity in Eastern Himalayan region of India. In: Goel R, Soni R, Suyal DC (eds) Microbiological advancements for higher altitude agro-ecosystems & sustainability. Springer, Singapore. https://doi.org/10.1007/978-981-15-1902-4_5
17. Behera RN, Nayak DK, Andersen P, Måren IE (2016) From jhum to broom: agricultural land-use change and food security implications on the Meghalaya Plateau, India. Ambio 45:63–77. https://doi.org/10.1007/s13280-015-0691-3
18. Mishra G, Giri K, Jangir A, Francaviglia R (2020) Projected trends of soil organic carbon stocks in Meghalaya state of Northeast Himalayas, India. Implications for a policy perspective. Sci Total Environ 698:134266. https://doi.org/10.1016/j.scitotenv.2019.134266
19. Teegalapalli K, Mailappa AS, Lyngdoh N, Lawrence D (2018) Recovery of soil macronutrients following shifting cultivation and ethnopedology of the Adi community in the Eastern Himalaya. Soil Use Manage 34:249–257. https://doi.org/10.1111/sum.12420
20. Bhalerao AK, Kumar B, Singha AK, Jat PC, Bordoloi R, Deka Bidyut C (2015) West Garo Hills district inventory of agriculture. ICAR-Agricultural Technology Application Research Institute, Umiam, Meghalaya, India

21. Soil Survey Staff (2014) Keys to soil taxonomy, 12th edn. United States Department of Agriculture, Natural Resources Conservation Service, Washington, DC, USA
22. Walkley A, Black IA (1934) An examination of the Degtjareff method for determining soil organic matter and a proposed modification of the chromic acid titration method. Soil Sci 37:29–38
23. Rich CI (1965) Elemental analysis by flame photometry. In: Black CA (ed) Methods of soil analysis, part 2: chemical and microbiological properties. American Society of Agronomy, Inc., Madison, WI, USA, pp 849–864. https://doi.org/10.2134/agronmonogr9.2.c3
24. Schollenberger CJ, Simon RH (1945) Determination of exchange capacity and exchangeable bases in soil—ammonium acetate method. Soil Sci 59:13–24
25. Moran EF, Brondizio ES, Tucker JM, da Silva-Forsberg MC, McCracken S, Falesi I (2000) Effects of soil fertility and land use on forest succession in Amazônia. For Ecol Manage 139:93–108. https://doi.org/10.1016/S0378-1127(99)00337-0
26. Lu D, Moran E, Mausel P (2002) Linking Amazonian secondary succession forest growth to soil properties. Land Degrad Dev 13:331–343. https://doi.org/10.1002/ldr.516
27. Mishra G, Jangir A, Francaviglia R (2019) Modeling soil organic carbon dynamics under shifting cultivation and forests using RothC model. Ecol Model 396:33–41. https://doi.org/10.1016/j.ecolmodel.2019.01.016
28. Bhattacharyya T, Chandran P, Ray SK, Mandal C, Pal DK, Venugopalan MV, Durge SL, Srivastava P, Dubey PN, Kamble GK, Sharma RP, Wani SP, Rego TJ, Pathak P, Ramesh V, Manna MC, Sahrawat KL (2007) Physical and chemical properties of selected benchmark spots for carbon sequestration studies in semi-arid tropics of India, global theme on agroecosystems report no. 35. International Crops Research Institute for the Semi-Arid Tropics, Patancheru, Andhra Pradesh, India
29. Hajabbasi MA, Jalalian A, Karimzadeh HR (1997) Deforestation effects on soil physical and chemical properties, Lordegan, Iran. Plant Soil 190:301–308. https://doi.org/10.1023/A:1004243702208
30. Lonsdale WM (1988) Predicting the amount of litterfall in forests of the world. Ann Bot 61:319–324. https://doi.org/10.1093/oxfordjournals.aob.a087560
31. Kalita RM (2015) Carbon storage and sequestration in tea agroforestry system in Barak Valley Assam. PhD thesis. Assam University, Silchar, Assam, India
32. Pal S, Panwar P, Bhardwaj DR (2013) Soil quality under forest compared to other land-uses in acid soil of north western Himalaya. India Ann For Res 56:187–198
33. Dutta D, Bandyopadhyay S, Baruah U, Sarkar D (2013) Use of taxonomic approach to estimate total soil organic carbon (SOC) pool sizes in Meghalaya state, India. J Indian Soc Soil Sci 61:237–243
34. Lantz A, Lal R, Kimble JM (2002) Land use effects on soil carbon pools in two major land resource areas of Ohio, USA. In: Kimble JM, Lal R, Follett RF (eds) Agricultural practices and policies for carbon sequestration in soil. CRC Press, Boca Raton, FL, USA, pp 165–176

Agronomic Aspects of Nutrient Management in Shifting Cultivation Agro-Ecosystems of North-East India

Sanjay Kumar Ray, Dibyendu Chatterjee, Saikat Ranjan Das, Sanjib Ray, Basant Kumar Kandpal, and Vinay Kumar Mishra

Abstract Shifting cultivation (locally called jhum) is a predominant form of farming in North-east India. In this system, slashing and burning are used to clear land in order to prepare it for shifting cultivation. After one to two years of cultivation, the land is then left fallow to allow for natural regrowth. In recent times, reduction in the fallow length has impacted the environment, including forest regeneration, resulting in continuous exploitation of natural resources and declining sustainability. Traditional shifting cultivation methods can be improved by promoting stress-tolerant rice and maize varieties, replacing slash-and-burn with the slash-and-char method, inclusion of legume crops, recycling of crop residues and in-situ residue management, nutrient management through balanced fertiliser use, harvesting of rainwater, establishing orchards, and using better fallow management techniques. Some constraints and challenges in nutrient management of jhum cultivation systems include the traditional farmers' mindset, undulating topography and soil erosion, non-application of chemical fertilisers, residue burning, and insufficient facilities for procurement of agri-inputs. Despite that, huge numbers of tribal farmers are still involved in the system due to their cultural heritage and lack of alternative employment opportunities to sustain their livelihood needs throughout the year.

Keywords Jhum rehabilitation · Nutrient loss · Organic nutrients · Soil erosion

---

S. K. Ray (✉) · B. K. Kandpal
ICAR Research Complex for NEH Region, Tripura Centre, Lembucherra, Tripura 799210, India
e-mail: sanjayray2006@gmail.com

D. Chatterjee · S. R. Das
ICAR National Rice Research Institute, Cuttack, Odisha 753006, India

S. Ray
Department of Agriculture, Government of West Bengal, Bamongola, Malda, West Bengal 732147, India

V. K. Mishra
ICAR Research Complex for NEH Region, Umiam, Meghalaya 793103, India
e-mail: vk.mishra@icar.gov.in

S. K. Tripathi and F. Q. Brearley (eds.), *Shifting Cultivation Systems*,
https://doi.org/10.1007/978-3-031-70388-1_10

1 Introduction

Shifting cultivation, also known as slash-and-burn, is a traditional farming method that has always been used in the hilly areas of North-east India [1, 2]. Shifting cultivation entails cutting and burning forests on hillsides at altitudes up to around 1500 m, cultivating for a year or two, and then abandoning the land to allow secondary forests to regrow and regenerate soil fertility. This method follows a typical cycle of 10–15 years, which ensures that the forests grow back adequately. In North-east India, 1.67 million hectares are under shifting cultivation [3] and indigenous people in most of the North-eastern mountainous states practice slash-and-burn, known locally as jhum [1, 4]. Furthermore, it is a ritually sanctioned way of life, a measure of self-identity deeply rooted in the ethos, culture, and religious aspirations of the hill tribal communities, apart from its direct contribution to food security and livelihoods in isolated hill areas [5].

Environmental systems such as regional hydrology, carbon (C) stocks, forest fires, and the provision of other ecological services are affected by forest clearing during shifting cultivation [1, 4, 6]. The impacts of frequent slash-and-burn agriculture and shorter agricultural cycles have been exacerbated by increasing populations [1, 2, 7]. Originally, when shifting cultivation emerged, it followed a cycle of 15–20 years that allowed vegetation to extract sufficient nutrients from the subsoil and enrich the surface soil by decomposing litterfall. However, the North-eastern region has experienced rapid population growth in recent decades, resulting in increased agriculture on marginal land and a reduced jhum cycle to up to just five years in length, leaving less time for naturally regenerating soil fertility. Farmers of the North-eastern region of India rarely use fertilisers or manures to produce their crops [8] and the use of agrochemicals such as insecticides is very low [9], hence the shifting cultivation is classified as 'organic by default' [10].

In comparison to other parts of India, the North-east has a relatively high soil organic carbon (SOC) content [11, 12] along with favourable agro-climatic conditions, livestock management in homeyards, and rich indigenous technical knowledge that are advantageous for the promotion of organic farming in the region [13, 14]. Nevertheless, the majority of tribal farmers participate in the system because they believe in their cultural heritage, and lack alternative means to meet their family's needs. This chapter presents several indigenous and potentially improved practices of nutrient management in shifting cultivation for sustainable farming practices to increase economic output with minimal negative impacts on the soil and the broader environment.

2 Evolution of Shifting Cultivation

Upland farmers experience numerous difficulties in meeting their daily food requirements in the hard to manage, remote, mountainous areas, which are generally unsuitable for the establishment of agro-ecosystems. The earning of a livelihood by jhumias (farmers involved in shifting cultivation) from shifting cultivation is not a 'choice', rather a 'compulsion' due to the hardships faced by the hill communities to fulfil their daily food requirements. Jhumias typically practice mixed farming, with the crop mix varying from tribe to tribe within a region. Rice (*Oryza sativa*; Poaceae), maize (*Zea mays*; Poaceae), millet (*Eleusine coracana*; Poaceae) and Job's tears (*Coix lacryma-jobi*; Poaceae) are the principal food grains, while cotton (*Gossypium arboreum*; Malvaceae), ginger (*Zingiber officinale*; Zingiberaceae), pigeon pea (*Cajanus cajan*; Fabaceae), pineapple (*Ananas comosus*; Bromeliaceae), rapeseed (*Brassica rapa*; Brassicaceae), and sesame (*Sesamum indicum*; Pedaliaceae) are key cash crops, with vegetables grown in jhum fields including cassava (*Manihot esculenta*; Euphorbiaceae), chillies (*Capsicum* sp.; Solanaceae), cucumbers (*Cucumis sativus*; Cucurbitaceae), onion (*Allium cepa*; Amaryllidaceae), potato (*Solanum tuberosum*; Solanaceae), pumpkins (*Cucurbita pepo*; Cucurbitaceae), rajma bean (*Phaseolus vulgaris*; Fabaceae), soybean (*Glycine max*; Fabaceae), taro (*Colocasia esculenta;* Araceae), and yams (*Dioscorea* sp. Dioscoreaceae) [15]. These crops are harvested at different times, providing the farmers with a diverse diet for almost six to nine months of the year [16]. In addition, soil-depleting crops such as cotton, maize, millet, rice, and soil-enriching crops such as legumes are grown together. This approach provides a number of direct and indirect benefits.

Both men and women are involved in shifting cultivation practices, but women bear more responsibility, particularly for field preparation, sowing, weeding, harvesting, husking, thrashing, and grain storage. In view of the fact that jhum fulfils their consumption needs for the entire year, the farming can be termed as self-sufficient. However, some farmers who are economically better off grow cash crops such as ginger, large cardamom (*Amomum subulatum*; Zingiberacaee), and turmeric (*Curcuma longa*; Zingiberaceae), and fruits such as apple (*Malus domestica*; Rosaceae), kiwi (*Actinidia deliciosa*; Actinidiaceae), orange (*Citrus* × *sinensis*; Rutaceae), and passion fruit (*Passiflora* sp.; Passifloraceae).

3 Impacts of Shifting Cultivation

Shifting cultivation enables tribal families to preserve their rich cultural traditions and diversity. Jhumias conserve agricultural biodiversity, mainly indigenous food crops such as rice, various vegetables, and even diverse fruits, in addition to forest biodiversity [17]. During the site selection process for shifting cultivation, jhumias take care to avoid areas dominated by rare or medicinally important species. The jhumias do not burn areas carelessly because they understand the process of fire-fighting and they remain awake all night to avoid fire spreading during the burning. They normally grow eight to 30 different species in a single jhum field [14, 16] with the complementarity allowing them to obtain more food items within a given period. Jhum is an integrated approach to supporting agroecosystems in the difficult topographies by incorporating natural forests, biodiversity and soil, and by following indigenous traditional knowledge. After clearing the forest, dried trees and branches are stored as fuel wood and the remaining residues are burned to clear the field during preparation for shifting cultivation. Ash from dried leaves and small branches provide a source of nutrients for crop growth. The practice of shifting cultivation is organic, using only agricultural residues, animal wastes etc. As a result, jhumias' minimal use of synthetic fertilisers may help to protect soil health to some extent. Jhumias have a unique approach to pest management whereby medium-sized plants are pruned to a specific height based on the expected height of the crop. During the early growth stages of the plant, particularly rice, the stumps provide a platform for birds to sit and feed on insects in the leaves of the rice, but as the rice grow to maturity, the stumps grow out and prevent the birds from eating the grains [18]. Minimal soil disturbance from no-tillage practices and placement of half burned logs across the slopes of the field also reduces soil erosion and surface runoff [18].

The impacts of shifting cultivation on soil erosion and land degradation have been established by numerous researchers. According to studies [19, 20], soil losses in shifting agriculture systems with slopes of 10° to 25° ranged from 6.3 to 230 Mg ha^{-1} yr^{-1}. In other studies, soil loss during the first and second years was estimated to be 70.1 to 149 Mg ha^{-1} yr^{-1} and 50.7 to 85.7 Mg ha^{-1} yr^{-1}, respectively, with corresponding available NPK nutrient losses of 24.6–27.0 kg ha^{-1} yr^{-1} and 15.9–17.8 kg ha^{-1} yr^{-1} [21, 22]. Reduction of the jhum cycle from 20 to 30 years to 3 to 4 years exacerbated top soil erosion (16.4 Mg ha^{-1} yr^{-1}) and nutrient loss, resulted in declining soil fertility and cereal crop productivity under shifting agriculture systems in Mizoram [23, 24]. Increased soil erosion is invariably linked to greater SOC loss [25, 26]. Comparative studies of the dynamics of SOC stocks between jhum cultivated cropland and undisturbed forest stands found that the decline in SOC under jhum is substantial in the short term (one to five years) but the losses stabilised in the range of five to ten years [27].

4 Area Under Shifting Cultivation and Its Productivity

In North-east India, the total area under shifting cultivation is estimated to be 0.9×10^6 ha, including both current jhum and abandoned jhum, accounting for 85% of the entire shifting cultivated land in India [28]. Moreover, with increases in population, the average size of land holdings are gradually being reduced. State-wise per family jhum field area is greatest in Arunachal Pradesh (1.29 ha) and Manipur (1.29 ha) followed by Mizoram (1.26 ha), Assam (1.20 ha), Meghalaya (1.01 ha), Tripura

(0.51 ha) and least in Nagaland (0.16 ha) [3]. This is partly because more than two-thirds of the total geographical area of the region is mountainous and extensive areas are not available for cultivation. Productivity of different crops is presented on a state-wise basis in Table 1.

5 Improved Management Strategies

The following measures can be considered to improve nutrient management and soil and water conservation under shifting agriculture in North-east India.

5.1 Promotion of Stress-Tolerant Varieties

Traditional crop varieties either do not respond, or respond poorly, to fertilisers [31]. Farmers are also reluctant to apply fertilisers to jhum fields [31, 32], so there is little scope for replacing these varieties with fertiliser-responsive high-yielding crop varieties. Therefore, stress-tolerant rice and maize varieties (e.g., cold stress, drought stress and iron-toxicity stress) and other stress-tolerant crop varieties can be promoted in shifting cultivation practices [30].

5.2 Replacement of Slash-and-Burn System with Slash-and-Char Method

To improve the soil quality of jhum fields, the slash-and-burn method can be replaced with slash-and-char [33], which is a C and nutrient conserving alternative to the existing slash-and-burn practice. Carbon is preserved in this technique compared to slash-and-burn since only biomass from the same farming area is used to make the charcoal [31, 33] Existing slash-and-burn systems degrade soil over time and produce greenhouse gases, and there may be opportunities to improve this system by switching to slash-and-char systems. If slash-and-burn is substituted with slash-and-char, up to 12% of total anthropogenic C emissions via land-use change can be offset in soil [31, 34]. Therefore, charcoal production for soil amelioration could create a C sink and be an important step toward sustainability and soil organic matter conservation in shifting cultivation systems.

Table 1 Productivity of major crops growing under shifting cultivation in the seven states of North-east India

Crop	Latin name	Productivity (Mg ha^{-1} yr^{-1})						
		Arunachal Pradesh	Assam	Manipur	Meghalaya	Mizoram	Nagaland	Tripura
Rice	*Oryza sativa*	1.14	1.50	1.34	1.60	1.21	1.91	1.07
Maize	*Zea mays*	2.60	–	2.41	1.35	1.54	1.95	1.32
Job's tears	*Coix lacryma-jobi*	–	–	–	–	–	1.12	–
Sesame	*Sesamum indicum*	–	–	–	–	1.14	1.66	0.59
Groundnut	*Arachis hypogaea*	–	–	1.24	–	–	1.00	1.09
Soybean	*Glycine max*	2.10	–	1.66	–	–	1.22	–
Pigeon pea	*Cajanus cajan*	–	–	1.20	–	–	1.10	0.72
Ricebean	*Vigna umbellata*	–	–	1.30	–	1.34	1.09	–
Potato	*Solanum tuberosum*	–	–	1.40	–	4.61	1.95	–
Cassava	*Manihot esculenta*	–	–	–	–	–	8.18	–
Taro	*Colocasia esculenta*	–	–	–	17.2	–	12.0	–
Ginger	*Zingiber officinale*	–	–	–	8.60	7.34	7.50	8.29
Source(s)		[18, 72]	[18]	[29]	[14]	[24]	[30]	[16]

5.3 Adoption of Good Agronomic Practices

Potential limitations of the jhum cultivation system are the use of local varieties, lack of fertiliser or manure application, and a lack of pest and disease control. Therefore, adoption of crop establishment methods such as using adequate quantities of seeds, sowing in rows across the slope with spacing of 25 cm, applying 50% of the recommended dose of fertilisers from organic or inorganic sources, and selecting suitable crop varieties could ensure long-term production and income for poor and marginal jhum farmers.

5.3.1 Rice

Improved agronomic management practices, such as half-doses of recommended fertiliser, half-doses of recommended NPK with farmyard manure (FYM), half-doses of recommended fertiliser for rice and intercropping with groundnut *Arachis hypogaea* (Fabaceae) and/or soybean at a 2:1 ratio, and two foliar applications of ammonium phosphate at 30 and 60 days after sowing increased average jhum rice yield by 68% and 116% for local and improved rice, respectively, when compared to traditional shifting cultivation practices [35]. Cultivation of jhum rice with management approaches such as micro-dosing of NPK and seed bio-priming with *Azotobacter* and phosphate-solubilising bacteria biofertilisers produced 96–109% greater grain yields than traditional practices [29]. Rice equivalent yield increased by 25–62% when jhum rice was grown in a strip-cropping system with soybean and groundnut, compared to traditional jhum (2.63 Mg ha^{-1}) farming [29]. Sowing of soybean or groundnut in 30 cm lines with rice in a 1:2 ratio helped to improve soil fertility and reduced soil loss in jhum fields [35]. Hence, efficient production strategies, such as strip-cropping cereals with legumes, can ensure efficient resource usage and enhance production [36, 37]. Improved approaches through the use of need-based fertilisers and biofertilisers also increased jhum rice yield in Tripura [16].

5.3.2 Maize

After rice, maize is the second most important crop in jhum cultivation. The application of biofertilisers, and micro-dosing (small doses of fertiliser-such as a "three-finger pinch" are placed in the planting hole at the time of sowing or at the base of the plants two weeks later) of NPK (20:20:20) at a rate of 3 g $plant^{-1}$ increased maize production by 53–62% as compared to traditional jhum maize cultivation [29]. Bio-priming of seeds with biofertilisers assisted mineralisation of organic matter in soil by microbes through enzymatic processes that convert it into available plant nutrients and other plant growth promoting compounds [38] resulting in increased maize growth and productivity.

5.3.3 Maize-Soybean Cropping Sequence

The maize equivalent yield in shifting cultivation system of Mizoram is around 1.5 Mg ha^{-1}. Introduction of high-quality protein maize in a maize-soybean cropping sequence enhanced maize equivalent yield by 4.6 Mg ha^{-1} with an increase in net income of Rs. 90,000 ha^{-1} [24]. Furthermore, planting of soybean after harvesting maize (i.e., in the first week of August) increased cropping intensity with an additional soybean yield of 1.48 Mg ha^{-1} as a second crop harvest. Thus, the adoption of soybean as a second crop might provide an additional net return of Rs. 50,000 ha^{-1} [24].

5.3.4 Ginger

The performance of ginger in raised beds was better in terms of rhizome yield as compared to flat-bed ginger cultivation [39]. Application of 120 kg nitrogen (N) ha^{-1} in a raised bed system by using FYM and vermicompost at equal doses yielded 12.9 Mg ha^{-1}of ginger rhizomes as compared to typical farmers' practice (FYM application at 5 Mg ha^{-1}) that yielded 8.6 Mg ha^{-1}. According to a nutrient management study of ginger, cultivation of soybean as a green manure crop in the inter-row spaces of the ginger and incorporation during the earthing-up operation is of great benefit in biomass production of ginger [39].

5.4 Integrated Soil Fertility Management

Farmers are often reluctant to apply fertilisers to jhum fields [13]; therefore, high yielding varieties will be unable to provide the desired yield. Furthermore, fertiliser use efficiency will be low, as most of the applied fertilisers will be lost to surface runoff due to the region's high rainfall [31]. Therefore, nutrient management through conventional organic sources of nutrients (compost, vermicomposts, FYM etc.) should be prioritised. Maintenance of greater soil organic matter in jhum field will reduce soil erosion and help crops to overcome moisture stress, especially during the dry winter months. According to studies, it was found that application of fertilisers or a combined application of fertiliser and FYM at 50% of the recommended dose could enhance yield of jhum rice by 40–60% [35]. Intercropping of legumes with cereals (rice and maize) and seed inoculation with biofertilisers improved root growth, biomass, and legume symbiotic N fixation, thereby enhancing soil fertility [29, 40]. Foliar application of nutrients (diammonium phosphate) twice (30 and 60 days after sowing) to jhum rice enhanced yield by 20–35% irrespective of the variety, in comparison with farmers' usual practice (where no manure or fertiliser was utilised) [35]. Biofertilisers can also play an important role in regulating soil nutrients in jhum farming systems, therefore microbial biodiversity should be further exploited for production of biofertilisers [41, 42].

5.5 Introduction of Legume Crops in the Rotation

Incorporating leguminous cover crops in a shifting cultivation system could help to reduce soil loss, improve soil health, suppress weeds, and provide food and nutrition for humans and animals, besides increasing incomes. In shifting cultivation systems of Manipur, promotion of pulse crops such as lentil (*Vicia lens*; Fabaceae), peas (*Pisum sativum*; Fabaceae), pigeon pea, rajma bean, and ricebean (*Vigna umbellata*; Fabaceae) yielded 1.4–1.8 Mg pea ha^{-1}, 1.2–1.8 Mg of pigeon pea ha^{-1}, 1.4–1.9 Mg rajma bean ha^{-1}, and 1.3–1.7 Mg ricebean ha^{-1} [43]. As jhumias follow a mixed cropping system there is considerable scope for promotion of pulse crops in shifting cultivation systems. Prakash et al. [44] suggested that broad bean (*Vicia faba*), cowpea (*Vigna unguiculata*), lima bean (*Phaseolus lunatus*), mungbean (*Vigna radiata*), peas, pigeon pea, ricebean, soybean, tree beans (*Parkia timoriana*) urdbean (*Vigna mungo*), winged bean (*Psophocarpus tetragonolobus*; all Fabaceae) are all possible pulse crops, which can be grown as a single crop or intercropping as a part in a shifting cultivation system.

5.6 Recycling of Crop Residues and In-Situ Residue Management

Low N and very low phosphorus (P) content in soil, coupled with farmers' indifference towards fertiliser use are the major constraints affecting crop yields in this region of India. Burning of crop residue reduces the amount of biomass available for composting in shifting cultivation system thus collection of biomass for composting from surrounding forests can be considered. Pits near jhum fields are essential for re-using biomass from crop and non-crop areas to make composts [17]. According to Ray et al. [45], the establishment of a vermicompost structure of 4 m × 1.25 m × 1 m allowed for the recycling of 3.17 Mg of biomass annually, resulting in an average of 1.24 Mg of mature compost, containing equivalent to 53.5 kg of NPK nutrients.

A four-year study on in-situ residue management in cropping systems based on rice and vegetables such as cabbage (*Brassica oleracea*; Brassicaceae), carrot (*Daucus carota*; Apiaceae), potato, rajma bean, and tomato (*Solanum lycopersicum*; Solanaceae) was conducted where only the economically valuable parts of the crops were harvested and the remaining parts, including weed residues, were chopped and incorporated into the soil. The greatest amount of NPK recycling was found in the rice-potato cropping sequence. With the exception of rice-cabbage, residual soil fertility in terms of available NPK was found to be consistent among all crop sequences. As compared to inorganically treated plots, soil biological properties such as *Rhizobium* populations, phosphate-solubilising microorganisms, and earthworm activity were significantly greater in the plots with residue recycling [13]. No tillage combined with maize stalk retention and application of *Ambrosia* spp. (Asteraceae) mulch at a rate of 5 Mg ha^{-1} resulted in the greatest improvement in soil quality measures and enhanced output in maize-based cropping sequences [8]. Therefore, recycling of residues through composting techniques or in-situ application can be promoted in shifting agricultural systems.

5.7 *Balanced Use of Fertiliser Based on Soil Test Results*

Soil pH, SOC, and available nutrients vary widely between and within North-east Indian states (Table 2), therefore soil testing is critical for balanced fertilisation of different nutrient elements based on crop requirements. The aim of balanced fertilisation is to correct intrinsic soil nutrient deficiencies, increase crop yield, improve product quality, increase farm income, improve nutrient-use efficiency, improve soil fertility, avoid environmental damage, and restore soil fertility and productivity (regardless of fertiliser sources). Unbalanced fertiliser application leads to soil degradation as opposed to soil building through balanced fertilisation, and ultimately leads to soil degradation. Therefore, balanced fertilisation would assure good yield in accordance with crop nutrient requirements.

The efficiency of fertiliser or manures can be increased to some extent by integrated use of organic manure, split application of N and potassium fertilisers, band placement of N, granular urea, foliar application of nutrients, etc. In the soil–plant system, nutrient components of fertiliser behave differently. Some fertilisers, such as P, sulphur, and zinc, have a significant residual effect, and the standing crops take up only a fraction of the total applied amount. In such cases, it is not cost-effective to apply a full dose of fertiliser to each crop, particularly for single cropping sequences and double cropping sequences when residual nutrient availability should be considered during fertiliser application for the second crops.

Table 2 Nutrient availability in soils (range) influenced by shifting cultivation North-east Indian states

Parameter	Arunachal Pradesh	Manipur	Meghalaya	Mizoram	Nagaland	Tripura
pH	4.40–7.30	4.00–5.40	4.76–5.20	5.02–5.95	3.66–6.69	4.50–5.10
SOC (%)	0.6–5.6	0.9–3.9	0.3–3.4	2.1–2.6	1.0–3.0	0.6–0.8
N (mg kg^{-1})	53–200	50–83	60–252	23–300	70–147	228–263
P (mg kg^{-1})	3.8–7.5	4.7–12.8	2.0–6.8	2.4–8.0	3.5–8.4	1.5–6.3
K (mg kg^{-1})	73.6–146	37.5–78.8	47.8–131	65.6–213	20.5–200	67–171
Reference(s)	[46]	[29, 47]	[48, 49]	[50]	[51, 52]	[16]

5.8 Adoption of Soil and Water Conservation Measures

The permissible soil loss from agricultural land, also known as the tolerance limit, ranges from 7.5 to 12 Mg ha^{-1} yr^{-1} in different physiographic zones of India [53]. In Nagaland, the soil loss tolerance limit (> 11.2 Mg ha^{-1} yr^{-1}) [54] was exceeded on 1,380,000 ha (84% of the total geographical area; TGA). Similarly, 78% of the TGA in Meghalaya (1,747,200 ha), and 68% of the TGA in Arunachal Pradesh (5,656,700 ha) exceeded the soil loss tolerance limit [55]. Therefore, the region loses a considerable amount of nutrients through soil erosion each year with losses exacerbated by heavy rainfall. Soil loss can be reduced effectively by using soil conservation techniques such as contour trenches, contour bunds, vegetative bunds, grass waterways etc. [56]. According to research, farming methods integrating bench terrace agriculture and agri-horti-silvopastoral systems may successfully control runoff and soil losses. As these systems imitate natural forest, over 90–100% of annual rainfall can be retained in mixed land-use systems with soil water conservation methods [54]. Agroforestry and other mixed land-use systems were the most successful at conserving moisture and reducing peak runoff, making them the best methods for soil and water conservation in hilly ecosystems [54]. This might be attributed to the influence of high litterfall rates that may have boosted soil cohesion after decomposition in addition to tree roots that also strongly bind the lower soil horizons through their deep rooting systems [57].

Contour bunding and planting of *Tephrosia* spp. (Fabaceae) at 0.5–1.0 m vertical intervals on contours is beneficial to reduce soil loss and improve soil fertility on up to 25° slopes if the soil depth is more than 1.0 m. With a minimum 0.5 m soil depth, a livestock-based land-use system may be encouraged by implementing soil and water conservation measures such as contour bunds, trenches, and grassed streams [58]. Legumes and non-leguminous annuals and perennials, shrubs, and trees-based fodder crops can be chosen for year-round supply of fodder by guaranteeing soil fertility. For rearing of four or five animals, a one-hectare field with a *Setaria sphacelata* (Poaceae) and *Stylosanthes guianensis* (Fabaceae) (1:1) mixed fodder production system is appropriate [59]. By decreasing soil erosion and replenishing the soil with C and nutrients, the bio-restoration process helps to maintain productive potential of the soil. Zero tillage or minimum tillage, hedge cropping, mulching, cover cropping, and the use of biofertilisers and organic manure are examples of bio-based resource conservation strategies that have the potential to improve organic matter and soil health in the region [60].

Bio-terracing in jhum fields with fast-growing and leguminous hedgerow species such as *Cajanus cajan*, *Crotalaria juncea*, *Crotalaria tetragona*, *Flemingia macrophylla*, *Indigofera articulata*, and *Tephrosia candida* (all Fabaceae), has the potential to reduce soil loss and improve soil development. Hedgerows alone prevent 94% of soil erosion and 78% of runoff [61]. Mulching with twigs and stems of hedgerow plants retained 83% of the soil and 42% of the rainwater [31]. Soil loss was reduced by 22% when hedgerow plants were integrated into jhum plots as compared to soil loss (38.1 t ha^{-1} yr^{-1}) of traditional jhum sites in Nagaland [31, 61]. Therefore, contour hedgerows can be a simple and sustainable alternative farming technology on hill slopes.

5.9 Rainwater Harvesting

The North-eastern region of India receives abundant rainfall (> 2000 mm yr^{-1}). However, there is water scarcity during the pre- and post-monsoon seasons (November to March). Rainwater harvesting using water collection tanks is a viable solution for a variety of agricultural activities. It has been determined that the water collected in a 30 m^3 tank can cultivate 500 m^2 of land during the dry months of November to March, resulting in a 176–261% increase in net income over traditional farming approaches [14]. Installation of rainwater harvesting structures and establishment of integrated farming systems through agricultural intensification and diversification activities in shifting cultivation systems resulted in increased crop yield, cropping intensity, supplemented feed requirements for animal components, improved soil fertility, increased net income, and produced greater employment opportunities [45]. Similarly, water harvesting systems in Nagaland had positive impacts on livelihood improvement through diversification of farming activities (production of winter vegetables, fishery and piggery) during the dry winter season and was able to increase farm income by 200% over the previous years [30].

5.9.1 Location Specific Promotion of Improved Farming System Models

Integrated farming systems (IFS) are a production system practiced in the entire North-east region and the livestock/poultry component of IFS may serve as an important source of manure for jhum fields [62]. Furthermore, IFS is a widely accepted and sound technique for coordinating joint management of land, water, vegetation, livestock, and human resources at the same time [1]. Establishment of an IFS model in hilly areas could reduce the risk of land degradation, increase the production potential of the soil, and reduce the risk of environmental degradation. In Arunachal Pradesh, crops (maize, legumes, vegetables, spices, agroforestry trees), water harvesting structures, water conservation measures, vermicomposting, and piggery components are used in farming system modules such as rice-based jhum systems, Khasi mandarin (*Citrus × reticulata*; Rutaceae)-based cropping systems, banana (*Musa* spp.; Musaceae)-based cropping systems, pineapple-based cropping systems, and tuber crop-based cropping systems with the banana-based, taro-based, and pineapple-based cropping systems being the most profitable [18].

In terms of greater returns, year-round employment, and long-term livelihood sustainability, the IFS model with crops, horticulture, fishery, and livestock or poultry components proved beneficial in Nagaland [63]. Similarly, in the Lunglei district of Mizoram, the IFS model including vegetables, fishery, and piggery components was found to be beneficial in terms of net returns and benefit:cost ratio [64]. In Manipur, the promotion of IFS models with components of crops, vegetables, fruit, piggery, poultry, fishery, and water management, and introduction of improved cultivars, better performing breeds, and good management practices resulted in increased production, greater income, and year-round employment opportunities. Vegetable cultivation, fish farming, and pig farming were the most profitable IFS components with benefit:cost ratios of 7.4:1, 5.1:1, and 4.2:1, respectively. Farmers who integrated agricultural, livestock, fish, and horticulture on their farm achieved greater net income. Therefore, promotion of IFS and improved farming can help tribal families by increasing productivity, yield, nutritional requirements, and employment prospects [43].

Integrated paddy-fish farming, horticulture–poultry–fishery–vermicomposting farming, and fish-poultry-vermicomposting farming are all typical farming practices in Meghalaya. Among these farming systems horticulture–poultry–fishery–vermicomposting farming had the greatest net returns and benefit:cost ratio followed by rice-fish farming and fish-poultry farming, respectively [64]. Compared to conventional agriculture, promotion of IFS in Tripura by including vegetables, spices, poultry, piggery, apiculture and fishery could increase output by 18%, profitability by 49%, and employment generation by 289 person days ha^{-1} yr^{-1} [65]. Therefore, it can be considered that adoption of improved management practices in IFS with various crop and livestock components is an effective strategy to ensure long-term production, income generation, and employment opportunities for farmers engaged in shifting cultivation.

5.9.2 Establishment of New Orchards

The North-east region of India has immense potential to promote the growth of high-value horticultural crops; however, the benefits have not been realised due to a lack of land-use planning based on land resource information. Climatically, Arunachal Pradesh has a potential advantage with three distinct regions for growing temperate, subtropical and tropical fruits. Due to the climate, soils and interest of growers in orchard management, different zone-specific crops are grown [18].

According to Kalita et al. [18] organic cultivation of banana, citronella (*Cymbopogon nardus*; Poaceae), kiwi, large cardamom, orange, and pineapple are suitable alternatives in jhum improvement programmes in Arunachal Pradesh. Similarly, in Mizoram, cultivation of strawberry (*Fragaria* × *ananassa*; Rosaceae) using mulching technology is gaining popularity under shifting land-use systems after terracing the land. Strawberry yielded an average 1750 kg and offered a net income of Rs. 0.16 million per year from 0.1 ha of land, with a benefit:cost ratio of 2.6:1 [24]. Along with lemon (*Citrus* × *limon*; Rutaceae) and orange, banana is a preferred crop in most parts of Mizoram and people have a belief that planting banana will protect them from natural calamities and disasters, and it also provides them economic prosperity [66]. According to Rajkhowa et al. [30], cultivation of arecanut (*Areca catecu*; Arecaceae), banana, king chilli, kiwi, large cardamom, lemon, lychee (*Litchi chinensis*; Sapindaceae), orange, passion fruit, and pineapple for orchard development and as part of inter-cropping and multi-storied cropping system models are suitable alternatives in jhum improvement programmes in Nagaland. In consonance with land suitability evaluation, oil palm (*Elaeis guineensis*; Arecaceae), rubber (*Hevea brasiliensis*; Euphorbiaceae), and tea (*Camellia sinensis;* Theaceae), could be potential land-use options to replace rainfed upland rice in the jhum areas as per the recommendation of Ray et al. [22]. Moreover, tree bean-based [67] farming has the potential to generate greater household income in shifting cultivation systems compared to other agricultural activities [68]. Moreover, in Tripura, a cropping system based on tuber crops such as cassava, elephant-foot yam (*Amorphophallus paeoniifolius*: Araceae), sweet potatoes (*Ipomoea batatas*; Convolvulaceae) and yams with piggery and banana production has been found suitable [69]. In Meghalaya agri-horti-silvi-pastoral system, silvi-horticultural system and arecanut/palm-based farming systems were found to be promising in terms of yield and net returns [70].

6 Constraints to Nutrient Management

Some constraints and challenges in nutrient management of jhum cultivation systems are as follows: the traditional mindset of farmers, undulating topography and soil erosion, non-application of chemical fertilisers, residue burning, cultivation with traditional tools and implements, poor economic status and inadequate credit support leading to insufficient facilities for procurement of agri-inputs, etc. Therefore, NITI Aayog [4] recently recommended developing permanent farmland with the establishment of perennial crops, multiple cropping, crop rotations, mulching, and organic matter recycling. There is also a growing need to promote livelihood and income diversification in the region for marginalised farmers through strengthening agriculture extension programs. Some government and non-government programmes have recently facilitated support for livelihood diversification schemes by promoting enterprises such as dairy, poultry, and pig farming. It is now clear that the government intends to define land-use in such a way that jhumias will be able to obtain credit and agricultural benefits such as insurance and subsidies [71].

7 Conclusions

Shifting cultivation in North-east India is merely not a cultivation system, rather it is an established and culturally-valued means of livelihood. While it can be a sustainable and efficient method of producing food, it can also result in nutrient depletion and soil degradation over time. Nutrient management in shifting cultivation is thus a critical consideration for preserving soil fertility and ensuring long-term food production. There are several nutrient management strategies that can be used in shifting cultivation, including indigenous and improved nutrient management strategies. This includes the use of improved varieties, new crops in the mixed cropping system such as legumes, the use of organic inputs, crop recycling, and so on. The traditional farming attitude, the undulating topography and soil erosion, the non-application of pesticides and fertilisers, residue burning, the lack of adequate facilities for agri-input procurement, etc. are some limitations to successful nutrient management of shifting cultivation production systems. Although nutrient management is important in shifting cultivation, these limitations and difficulties make it challenging to put efficient nutrient management strategies into reality. A multifaceted strategy will be required to tackle these limitations and difficulties, including enhancing market access, increasing awareness and knowledge, improving resource access, and addressing the impacts of climate change. Although there are several constraints to managing shifting cultivation traditionally, improved nutrient management strategies could make shifting cultivation more sustainable in terms of yields and economic benefits. Advanced soil nutrient management may prove helpful as it could increase the shifting cultivation cycle, thus increasing its sustainability.

References

1. Kumar R, Patra MK, Thirugnanavel A, Chatterjee D, Deka BC (2016) Towards the natural resource management for resilient shifting cultivation system in Eastern Himalayas. In: Bisht J, Meena V, Mishra P, Pattanayak A (eds) Conservation agriculture: an approach to combat climate change in Indian Himalaya. Springer, Singapore, pp 409–436. https://doi.org/10.1007/978-981-10-2558-7_16
2. Mohapatra KP, Chandra P (2017) Rejuvenation of *jhum* land through agroforestry interventions. In: Prakash N, Roy SS, Ansari MA, Sharma SK, Punitha P, Sailo B, Singh IM (eds) Jhum improvement for sustaining farm livelihood and natural resource conservation in North Eastern hill region: vistas and frontiers. ICAR Research Complex for NEH Region, Umiam, Meghalaya, India, pp 73–88
3. Deka BC, Singha AK, Parisa D, Amrutha P (eds) (2020) Prospects of Northeast agriculture in post covid 19 scenario. ICAR-Agricultural Technology Application Research Institute, Umiam, Meghalaya, India
4. NITI Aayog (2018) Report of working group III: shifting cultivation: towards a transformational approach. New Delhi, India
5. Gupta V (2005) *Jhum* cultivation practices of the Bangnis (Nishis) of Arunachal Pradesh. Indian J Tradit Knowl 4:47–56
6. Metzger JP (2003) Effects of slash-and-burn fallow periods on landscape structure. Environ Conserv 30:325–333. https://doi.org/10.1017/S0376892903000341
7. Thong P, Pebam R, Sahoo UK (2018) A geospatial approach to understand the dynamics of shifting cultivation in Champhai district of Mizoram, North-East India. J Indian Soc Remote Sens 46:1713–1723. https://doi.org/10.1007/s12524-018-0832-9
8. Das A, Ghosh PK, Lal R, Saha R, Ngachan SV (2014) Soil quality effect of conservation practices in maize–rapeseed cropping system in Eastern Himalaya. Land Degrad Dev 28:1862–1874. https://doi.org/10.1002/ldr.2325
9. Patel DP, Das A, Kumar M, Munda GC, Ngachan SV, Ramkrushna G, Layek J, Pongla N, Buragohain J, Somireddy U (2015) Continuous application of organic amendments enhances soil health, produce quality and system productivity of vegetable-based cropping systems in

subtropical Eastern Himalayas. Exp Agric 51:85–106. https://doi.org/10.1017/S0014479714000167
10. Kuotsuo R, Chatterjee D, Deka BC, Kumar R, Ao M, Vikramjeet K (2014) Shifting cultivation: an 'organic like' farming in Nagaland. Indian J Hill Farming 27:23–28
11. Bhattacharyya T, Pal DK, Chandran P, Ray SK, Mandal C, Telpande B (2008) Soil carbon storage capacity as a tool to prioritize areas for carbon sequestration. Curr Sci 95:482–494
12. Choudhury BU, Mohapatra KP, Das A, Das PT, Nongkhlaw L, Fiyaz AR, Ngachan SV, Hazarika S, Rajkhowa DJ, Munda GC (2013) Spatial variability in distribution of organic carbon stocks in the soils of North East India. Curr Sci 104:604–614
13. Das A, Patel DP, Kumar M, Ramkrushna GI, Ngachan SV, Layek J, Lyngdoh M (2014) Influence of cropping systems and organic amendments on productivity and soil health at mid altitude of North East India. Indian J Agric Sci 84:1525–1530
14. Das A, Singh RK, Ramkrushna GI, Layek J, Tripathi AK, Ngachan SV, Choudhury BU, Patel DP, Rajkhowa DJ, Chakroborty D, Ghosh PK (2017) Roof water harvesting in hills—innovations for farm diversification and livelihood improvement. Curr Sci 113:292–298. https://doi.org/10.18520/cs/v113/i02/292-298
15. Bhuyan R (2019) A review note on shifting cultivation in Northeast India amidst changing perceptions. Dhaulagiri J Sociol Anthropol 13:90–95. https://doi.org/10.3126/dsaj.v13i0.24252
16. Kandpal BK, Bhowmik SN (2017) Shifting cultivation in Tripura: challenges, prospects and alternatives. In: Prakash N, Roy SS, Ansari MA, Sharma SK, Punitha P, Sailo B, Singh IM (eds) Jhum improvement for sustaining farm livelihood and natural resource conservation in North Eastern hill region: vistas and frontiers. ICAR Research Complex for NEH Region, Umiam, Meghalaya, India pp 57–72
17. Chatterjee D, Deka BC, Patra MK, Thirugnanavel A, Borah TR, Kuotsuo R, Ngachan SV (2012) Resilient shifting cultivation for sustainable soil-water-nutrient-plant continuum in hilly agriculture system of North Eastern India. In: Deka BC, Patra MK, Thirugnanavel A, Chatterjee D, Borah TR, Ngachan SV (eds) Resilient shifting cultivation: challenges and opportunities. ICAR Research Complex for NEH Region, Nagaland Centre, Jharnapani, Medziphema, Nagaland, India, pp 104–109
18. Kalita H, Baruah MS, Datta D, Jini D, Alone RA (2017) Status of shifting cultivation (*jhum*) in Arunachal Pradesh, India. In: Prakash N, Roy SS, Ansari MA, Sharma SK, Punitha P, Sailo B, Singh IM (eds) Jhum improvement for sustaining farm livelihood and natural resource conservation in North Eastern hill region: vistas and frontiers. ICAR Research Complex for NEH Region, Umiam, Meghalaya, India, pp 1–12
19. Pandey A, Mathur A, Mishra SK, Mal BC (2009) Soil erosion modelling of a Himalayan watershed using RS and GIS. Environ Earth Sci 59:399–410. https://doi.org/10.1007/s12665-009-0038-0
20. Poreba GJ, Prokop P (2011) Estimation of soil erosion on cultivated fields on the hilly Meghalaya plateau, North-East India. Geochronometria 38:77–84. https://doi.org/10.2478/s13386-011-0008-7
21. Chatterjee D, Kuotsu R, Ray SK, Patra MK, Thirugnanavel A, Kumar R, Borah TR, Chowdhury P, Pongen I, Satapathy BS, Deka BC (2022) Preventing soil degradation in shifting cultivation using integrated farming system models. Arch Agron Soil Sci 68:1841–1857. https://doi.org/10.1080/03650340.2021.1937139
22. Ray SK, Chatterjee D, Chowdhury P, Deka BC, Bihari P, Saha S (2021) Comparison of indigenous and mechanical conservation technologies for shifting cultivation agro-ecology of north-eastern Himalaya. J Indian Soc Soil Sci 69:60–71. https://doi.org/10.5958/0974-0228.2021.00021.9
23. Bhattacharyya R, Ghosh BN, Mishra PK, Mandal B, Rao CS, Sarkar D, Das K, Anil KS, Lalitha M, Hati KM, Franzluebbers AJ (2015) Soil degradation in India: challenges and potential solutions. Sustainability 7:3528–3570. https://doi.org/10.3390/su7043528
24. Singh SB, Boopathi T, Dutta SK, Singh AR, Lungmuana, Saha S, Dayal V (2017) Shifting cultivation system in Mizoram—status, determinants and strategies for viability and sustainability. In: Prakash N, Roy SS, Ansari MA, Sharma SK, Punitha P, Sailo B, Singh IM (eds)

Jhum improvement for sustaining farm livelihood and natural resource conservation in North Eastern hill region: vistas and frontiers. ICAR Research Complex for NEH Region, Umiam, Meghalaya, India, pp 33–41

25. Chatterjee D, Kuotsu R, Ao M, Saha S, Ray SK, Ngachan SV (2019) Does rise in temperature adversely affect soil fertility, carbon fractions, microbial biomass and enzyme activities under different land uses? Curr Sci 116:2044–2054. https://doi.org/10.18520/cs/v116/i12/2044-2054
26. Sahoo UK, Singh SL, Gogoi A, Kenye A, Sahoo SS (2019) Active and passive soil organic carbon pools as affected by different land use types in Mizoram, Northeast India. PLoS One 14:e0219969. https://doi.org/10.1371/journal.pone.0219969
27. Mishra G, Das J, Sulieman M (2019) Modelling soil cation exchange capacity in different land-use systems using artificial neural networks and multiple regression analysis. Curr Sci 116:2020–2027. https://doi.org/10.18520/cs/v116/i12/2020-2027
28. Ministry of Statistics and Programme Implementation (2014) Statistical year book India 2014. Ministry of Statistics and Programme Implementation. Government of India, New Delhi, India
29. Baishya LK, Ansari MA, Sarkar D, Ghosh M, Kumar S, Prakash N (2016) Productivity enhancement in shifting cultivated lands through biofertilizer and micro-dosing of NPK in Eastern Himalayan region. Res Crop 17:268–275. https://doi.org/10.5958/2348-7542.2016.00046.2
30. Rajkhowa DJ, Baishya LK, Ray SK, Sharma PR, Barman J, Ezung NK (2017) Challenges, scope, and opportunities of jhum rejuvenation in Nagaland. In: Prakash N, Roy SS, Ansari MA, Sharma SK, Punitha P, Sailo B, Singh IM (eds) Jhum improvement for sustaining farm livelihood and natural resource conservation in North Eastern hill region: vistas and frontiers. ICAR Research Complex for NEH Region, Umiam, Meghalaya, India, pp 42–56
31. Hazarika S (2017) Options for natural resources conservation for *jhum* improvement in NEH region. In: Prakash N, Roy SS, Ansari MA, Sharma SK, Punitha P, Sailo B, Singh IM (eds) Jhum improvement for sustaining farm livelihood and natural resource conservation in North Eastern hill region: vistas and frontiers. ICAR Research Complex for NEH Region, Umiam, Meghalaya, India, pp 103–109
32. Das A, Patel DP, Ramkrushna GI, Munda GC, Ngachan SV, Buragohain J, Kumar M, Naropongla (2013) Crop diversification, crop and energy productivity under raised and sunken beds: results from a seven-year study in a high rainfall organic production system. Biol Agric Hortic 30:73–87. https://doi.org/10.1080/01448765.2013.854709
33. Layek J, Narzari R, Hazarika S, Das A, Rangappa K, Devi S, Balusamy A, Saha S, Mandal S, Idapuganti RG, Babu S, Choudhury BU, Mishra VK (2022) Prospects of biochar for sustainable agriculture and carbon sequestration: an overview for Eastern Himalayas. Sustainability 14:6684. https://doi.org/10.3390/su14116684
34. Lehmann J, Gaunt J, Rondon M (2006) Bio-char sequestration in terrestrial ecosystems—a review. Mitig Adapt Strat Glob Change 11:403–427. https://doi.org/10.1007/s11027-005-9006-5
35. Layek J, Das A, Ramkrushna GI, Panwar AS, Verma BC, Roy A (2018) Improving rice production under shifting cultivation: a case study. In: Das A, Mohapatra KP, Ngachan SV, Panwar AS, Rajkhowa DJ, Ramakrushna GI, Layek J (eds) Conservation agriculture for advancing food security in changing climate (crop production, farming system and soil health), vol 1. Today & Tomorrow's Printers and Publishers, New Delhi India, pp 143–153
36. Ramakrishnan PS (2003) Linking natural resource management with sustainable development of traditional mountain societies. Trop Ecol 44:43–54
37. Jeeva SRDN, Laloo RC, Mishra BP (2006) Traditional agricultural practices in Meghalaya, North East India. Indian J Tradit Knowl 5:7–18
38. Burns RG, DeForest JL, Marxsen J, Sinsabaugh RL, Stromberger ME, Wallenstein MD, Weintraub MN, Zoppini A (2013) Soil enzymes in a changing environment: current knowledge and future directions. Soil Biol Biochem 58:216–234. https://doi.org/10.1016/j.soilbio.2012.11.009
39. Das A, Layek J, Babu S, Kumar M, Yadav GS, Patel DP, Idapuganti RG, Lal R, Buragohain J (2020) Influence of land configuration and organic sources of nutrient supply on productivity and quality of ginger (*Zingiber officinale* Rosc.) grown in Eastern Himalayas, India. Environ Sustain 3:59–67. https://doi.org/10.1007/s42398-020-00098-x

40. Chaturvedi S, Chandel AS, Dhyani VC, Singh AP (2010) Productivity, profitability and quality of soybean (*Glycine max*) and residual soil fertility as influenced by integrated nutrient management. Indian J Agron 55:133–137
41. Wapongnungsang, Ovung, EY, Upadhyay KK, Tripathi SK (2021) Soil fertility and rice productivity in shifting cultivation: impact of fallow lengths and soil amendments in Lengpui, Mizoram northeast India. Heliyon 7:e06834. https://doi.org/10.1016/j.heliyon.2021.e06834
42. Giri K, Mishra G, Suyal DC, Kumar N, Doley B, Das N, Baruah RC, Bhattacharyya R, Bora N (2023) Performance evaluation of native plant growth-promoting rhizobacteria for paddy yield enhancement in the jhum fields of Mokokchung, Nagaland, North East India. Heliyon 9:e14588. https://doi.org/10.1016/j.heliyon.2023.e14588
43. Ansari MA, Prakash N, Baishya LK, Punitha P, Sharma PK, Yadav JS, Kabuei GP, Levis KL (2014) Integrated farming system: an ideal approach for developing more economically and environmentally sustainable farming systems for the Eastern Himalayan Region. Indian J Agric Sci 84:356–362
44. Prakash N, Roy SS, Ansari MA, Sharma SK, Punitha P, Sailo B, Singh IM (eds) (2017) Jhum improvement for sustaining farm livelihood and natural resource conservation in North Eastern hill region: vistas and frontiers. ICAR Research Complex for NEH Region, Umiam, Meghalaya, India
45. Ray SK, Chatterjee D, Rajkhowa DJ, Baishya SK, Hazarika S, Paul S (2020) Effects of integrated farming system and rainwater harvesting on livelihood improvement in North-Eastern region of India compared to traditional shifting cultivation: evidence from an action research. Agrofor Syst 94:451–464. https://doi.org/10.1007/s10457-019-00406-3
46. Kumar J, Kalita H, Mukdoh B, Jini D, Alone RA, Oppo P (2020) Effect of different soil and water management practices on sustainability of jhum cultivation. J Pharmacogn Phytochem 9:1486–1488. https://doi.org/10.22271/phyto.2020.v9.i2x.11062
47. Sahoo S, Vasu D, Paul R, Sen TK, Ray SK, Chadran P (2020) Acid soils of Manipur of the North-eastern region of India: their mineralogy, pedology, taxonomy and edaphology. Clay Res 39:31–43. https://doi.org/10.5958/0974-4509.2020.00005.4
48. Sharma S, Choudhury P, Chetry N (2016) Status of soil nutrients and fertility of four land use system in Raid-Marwet region, Ribhoi district of Meghalaya, India. J Glob Biosci 5:4178–4187
49. Mishra VK, Saha R (2021) Characterization of soil health under different land use pattern in hilly eco-system of Meghalaya. http://www.kiran.nic.in/pdf/publications/Characterization_Soil_Health_under_Different_Land_Use_Pattern.pdf. Accessed 21 Mar 2023
50. Kenye A, Sahoo UK, Singh SL, Gogoi A (2019) Soil organic carbon stock of different land uses of Mizoram, Northeast India. AIMS Geosci 5:25–40. https://doi.org/10.3934/geosci.2019.1.25
51. Ray SK, Chatterjee D, Saha S, Kamble KH, Deka BC (2015) A guide for soil nutrient management with special reference to Wokha. ICAR Research Complex for NEH Region, Nagaland Centre, Jharnapani, Medziphema, Nagaland, India
52. Mishra G, Francaviglia R (2021) Land uses, altitude and texture effects on soil parameters. A comparative study in two districts of Nagaland, Northeast India. Agriculture 11:171. https://doi.org/10.3390/agriculture11020171
53. Lenka NK, Mandal D, Sudhishri S (2014) Permissible soil loss limits for different physiographic regions of West Bengal. Curr Sci 25:665–670
54. Saha R, Chaudhary RS, Somasundaram J (2012) Soil health management under hill agroecosystem of North East India. Appl Environ Soil Sci 696174. https://doi.org/10.1155/2012/696174
55. Chattopadhyay T, Bandyopadhyay S, Baruah U, Nayak D, Dhyani DL, Sarkar D, Singh SK (2014) Soil Erosion in Arunachal Pradesh. National Bureau of Soil Survey and Land Use Planning, Nagpur, Maharashtra, India
56. Chatterjee D (2018) Managing shifting cultivated areas for arresting land degradation. In: Das A, Mohapatra KP, Ngachan SV, Panwar AS, Rajkhowa DJ, Ramkrushna GI, Layek J (eds) Conservation Agriculture for Advancing Food Security in Changing Climate, vol 2. Today & Tomorrow's Printers and Publishers, New Delhi, India, pp 683–694

57. Saha R, Mishra VK, Tomar JMS (2005) Effect of agro-forestry systems on erodibility and hydraulic properties of Alfisols in eastern Himalayan region. Indian J Soil Conserv 33:251–253
58. Chatterjee D (2015) Integrated soil and water management in shifting cultivation. In: Rajkhowa, DJ, Das A, Ngachan SV, Sikka AK, Lyngdoh M (eds) Integrated soil and water resource management for livelihood and environmental security. ICAR Research Complex for NEH Region, Umiam, Meghalaya, India, pp 120–124
59. Verma ND, Satapathy KK, Singh RK, Singh JL, Dutta, KK (2001) Shifting cultivation and alternative farming systems. In: Verma ND, Bhatt BP (eds) Steps towards modernization of agriculture in NEH Region. ICAR Research Complex for NEH Region, Umiam, Meghalaya, India, pp 345–364
60. Yadav GS, Devi HL, Babu S, Das A, Meena RS (2020) Bio-restoration of degraded lands under the sub-tropical North Eastern Himalayan region of India. Indian J Hill Farm Spec Issue 18–27
61. Kumar R, Meena VS (2016) Towards the sustainable management of problematic soils in Northeast India. In: Bisht J, Meena V, Mishra P, Pattanayak A (eds) Conservation agriculture: an approach to combat climate change in Indian Himalaya. Springer, Singapore, pp 339–366. https://doi.org/10.1007/978-981-10-2558-7_13
62. Chatterjee D (2013) Model farming systems for resilient shifting cultivation in Nagaland. In: Chatterjee D, Ezung NK (eds) Resilient *jhum* cultivation through integrated farming system. ICAR Research Complex for NEH Region, Nagaland Centre, Jharnapani, Medziphema, Nagaland, India, pp 7–9
63. Kumar R, Patra MK, Thirugnanavel A, Deka BC, Chatterjee D, Borah TR, Rajesha G, Talang HD, Ray SK, Kumar M, Upadhayay PK (2018) Comparative evaluation of different integrated farming system models for small and marginal farmers under the Eastern Himalayas. Indian J Agric Sci 88 (11):1722–1729. https://doi.org/10.56093/ijas.v88i11.84913
64. Deka BC, Singha AK, Parisa D, Tariang AM, Kyndiah EK, Marak MR (eds) (2020) Integrated farming systems for doubling farmers' income in NEH region of India. ICAR-Agricultural Technology Application Research Institute, Umiam, Meghalaya, India
65. Sharmah D, Debnath B, Kandpal BK, Das D (2019) Profitability of Integrated Farming System model for small and medium farmers of South Tripura, Tripura. J Appl Nat Sci 11:587–589. https://doi.org/10.31018/jans.v11i3.2008
66. Sahoo UK, Vanlalhluna PC, Singh SL (2018) Indigenous technologies and local climate change adaptation practices around East Kawlchaw watershed, Saiha, Mizoram: a case study. In: Zothanzama J, Saitluanga BL, Lalnuntluanga, Lalzarzovi ST, Zonunsanga IR (eds) Climate change: impact, adaptation and response in the Eastern Himalayas. Excel India Publishers, New Delhi, India, pp 28–39
67. Ovung EY, Loya B, Brearley FQ, Tripathi SK (2021) Ethnic uses of *Parkia timoriana* (Fabaceae) and their significance to the Lotha tribes of Nagaland, North east India. Vegetos 34:77–85. https://doi.org/10.1007/s42535-020-00178-7
68. Rocky P, Sahoo UK, Thapa HS (2004) Livelihood generation through treebean (*Parkia roxburghii* G. Don) in Imphal West district of Manipur. J Non-Timber For Prod 11:135–139
69. Datta M, Singh NP (2012) Shifting cultivation: land degradation and an approach to remedial measures in North East-India. In: Devvarma NC (ed) Shifting cultivation in Tripura, Tribal Research and Cultural Institute, Government of Tripura, Argatala, India, pp 35–57
70. Jha AK, Rymbai H, Verma VK, Deshmukh NA, Talang H, Ruth Assumi S, Devi MB (2017) Role of horticulture in jhum improvement and rehabilitation. In: Prakash N, Roy SS, Ansari MA, Sharma SK, Punitha P, Sailo B, Singh IM (eds) Jhum improvement for sustaining farm livelihood and natural resource conservation in North Eastern hill region: vistas and frontiers. ICAR Research Complex for NEH Region, Umiam, Meghalaya, India, pp 89–103
71. Sharma Y (2020) Shifting cultivation may soon get legal stamp. https://economictimes.indiatimes.com/news/economy/agriculture/shifting-cultivation-may-soon-get-legal-stamp/articleshow/74200736.cms. Accessed 19 Mar 2023
72. Pattanayak A, Ngachan SV, Mohanty AK (2011) Status paper on rice in North East India. Rice Knowledge Management Portal, Directorate of Rice Research, Rajendranagar, Hyderabad, India

Shifting Cultivation to Settled Agriculture for Sustaining Soil Health and Agriculture Productivity in North-East India

Lungmuana, Christy B. K. Sangma, Thejangulie Angami, and Ingudam Shakuntala

Abstract Shifting cultivation or 'jhum' is a traditional system of crop cultivation following recovery of soil fertility via forest regeneration. The system is still prevalent in many parts of the world including the North-east of India. This subsistence system is embedded within traditional culture by many, yet there is a requirement for the improvement of the system for increased sustainability and economic productivity. Modern technologies, like application of fertiliser, biofertiliser, etc. proved to be beneficial for soil fertility improvement and crop productivity. In the recent past, there has been a tremendous change in land-use from traditional jhum cultivation to settled agriculture like agroforestry plantations in the North-east region of India. Land-uses that have been converted to plantations, with appropriate management, decrease soil degradation, conserve soil organic matter and soil health, besides improving farmers' livelihoods. Nevertheless, soil carbon sequestration can be achieved in abandoned jhum by multipurpose tree plantations on slopes where normal cultivation is difficult. Integrated farming systems (IFS), through scientific integration of agricultural and horticultural crops, trees, agroforestry, and animal components recycle plant litter, and conserve soil health and available natural resources to increase farmers' livelihoods. Thus, this chapter aims to highlight options from shifting cultivation to alternate land-uses to achieve sustainability in the hills of North-east India.

Keywords Land-use change · Mixed farming · Settled agriculture · Soil health · Traditional land-use

---

Lungmuana (✉) · I. Shakuntala
ICAR-RC-NEH Region, Mizoram Centre, Kolasib, Mizoram 796081, India
e-mail: Lungmuana@icar.gov.in

C. B. K. Sangma
ICAR-RC-NEH Region, Umiam, Meghalaya 793103, India

T. Angami
ICAR-RC-NEH Region, Arunachal Centre, Basar, Arunachal Pradesh 791101, India

I. Shakuntala
College of Veterinary Sciences and Animal Husbandry, Central Agricultural University, Jalukie, Peren, Nagaland 797110, India

S. K. Tripathi and F. Q. Brearley (eds.), *Shifting Cultivation Systems*,
https://doi.org/10.1007/978-3-031-70388-1_11

1 Introduction

Shifting cultivation, swidden, or slash-and-burn is a subsistence system of cultivation dating its origin to the Neolithic period (13,000 to 3000 BC) [1]. It has been widely practiced by groups in Asia, Africa, and Latin America providing livelihoods to millions of people over the world [2]. The practice is a prevalent form of agriculture in the tribal dominated North-east hills of India, where it is locally known as 'jhum'. The practice is carried out over about 1.47 million ha of land and provides livelihoods to 0.44 million people in the region [3]. Although the practice differs between locations, the main principle behind this system is the accumulation of organic matter and restoration of soil fertility through forest regeneration. Annual crops are cultivated for one to three year after burning the slashed vegetation, following which the site is left fallow for forest regeneration. There have been numerous reports on the length of fallow periods being shortened leading to reduced soil health [4]. At least a ten-year fallow period has been recommended to regain soil health [5, 6] and soil health may not be regained in shorter time frames. Because shifting cultivation is practiced under tropical humid conditions, soil health can be drastically reduced within one year of cultivation due to removal of soil nutrients through crop biomass harvest and runoff loss because of high rainfall and steep slopes [6]. To enhance the system for more sustainable approaches, strategies for soil fertility enrichment under the shortened fallow periods should be prioritised [7].

The practice of slash-and-burn cultivation mostly exists in tribal areas and is interlinked with the ecology, culture and economics of the communities. However, it has been observed that the practice has been changing slowly towards settled forms of agriculture. For example, in Mizoram, the State Government has launched the New Land Use Policy (NLUP) to transform the natural fallows to permanent plantations or agroforestry, paving a way for more economic reforms for the farmers. Agroforestry plantation systems have been reported to sequester carbon (C) both above- and below-ground and improve soil health [8, 9]. Nevertheless, conversion of forest to agricultural land is a common practice due to population pressure. Agroforestry plantation systems may be practiced for many years, while repeated shifting cultivation cycles may limit the ecosystem recovery by decreasing the organic matter inputs and further degrading soil health [10]. Soil organic matter and nutrient depletion, soil acidity, and exposure of acidic sub-soil due to erosion also have to be addressed. Thus, integrating plantations in hill agriculture may maintain the production capacity of agricultural systems as well as maintaining ecosystem services. In this chapter, we attempt to review the impact of shifting cultivation and its effect on soil health and ways to achieve soil sustainability through external inputs like fertiliser, the role of agroforestry plantations in enhancing soil health and C sequestration, whilst considering enhancing farmers' livelihoods in the hills of North-east India.

2 Shifting Cultivation and Soil Fertility Management

Soil organic matter accumulation through forest regeneration is one of the major drivers for land selection in shifting cultivation and for which long fallow periods to restore essential soil nutrients are important [11]. One of the earliest studies on soil fertility restoration by increase in biomass through forest regeneration in shifting cultivation was reported by Nye and Greenland [12]. In North-east India, various authors [5–7] reported that cultivation of older forest fallows leads to fields with more nutrients and reduced bulk density after burning. Although soil available nutrients are generally studied, soil organic carbon (SOC) turnover in relation to soil biological activities such as soil microbial biomass and soil enzyme activities also act as a sensitive indicator and rapidly respond to soil management and disturbance [13]. Lungmuana et al. [6] studied biochemical properties i.e. microbial biomass carbon and soil enzyme activities (phosphatases, dehydrogenase, arylsulphatase etc.) to determine the soil health status as affected by the length of fallow period and agricultural operations in the first two years of cultivation. In that study, fallow periods from three to twenty-three years were selected, and it was concluded that the fallow period should be at least ten years to conserve soil health.

From different studies, it is clear that fallow period length influences the increase in soil organic matter. However, as reported by many authors regarding the reduction in the length of the forest fallow periods [14–16], there is an urgent need for the application of modern technologies to sustain soil fertility and crop productivity in jhum fields. In the North-east Indian states, jhum sites are abandoned after the first year of cultivation. Thus, challenge lies in the utilisation of land beyond this first year and shortened forest fallows through external inputs to sustain crop yields. Rice (*Oryza sativa*; Poaceae) and maize (*Zea mays*; Poaceae) are the major crops. Several other species like perilla (*Perilla frutescens*; Lamiaceae), cow pea (*Vigna unguiculata*; Fabaceae), brinjal (*Solanum melongena*; Solanaceae), bitter gourd (*Momordica charantia*; Cucurbitaceae), ginger (*Zingiber officinale*; Zingiberaceae), and turmeric (*Curcuma longa*; Zingiberaceae), etc. are grown as minor inter-crops. A result of field experiments conducted in Manipur comparing improved jhum management practice (seed bio-priming with microbial inocula such as *Azotobacter* + phosphate-solubilising bacteria, and micro-dosing of NPK) and traditional jhum, showed that improved management increased the soil fertility and crop productivity [17]. Additionally, legume intercropping with groundnut (*Arachis hypogaea*; Fabaceae) and soybean (*Glycine max*; Fabaceae) also increased soil fertility [17]. Mulching or residue incorporation can also be very effective in improving soil properties and SOC as well as increasing crop productivity and economic returns [18].

Wapongnungsang et al. [16] studied the application of rhizosphere microbial inocula and rock phosphate on the soil physicochemical properties and yield of rice in fields derived from a period of ten- and 15-year-old jhum fallows in Mizoram. Results from this study suggested that there was a synergistic effect of microbial inocula and rock phosphate in increasing rice productivity and soil fertility. Another study was from three-year-old fallows where chemical fertilisers were used on maize

[19]. In this study, three local maize landraces, namely *mimpui* flint corn, *mimban* sticky maize and *puakzo* popcorn, were grown with fertiliser application at a dose of 80:60:40 kg ha^{-1} through urea, single super phosphate and muriate of potash to the jhum slope (10°) via dibbler application. The initial soil fertility was poor (pH: 5.2; SOC: 0.84%; Available N, P and K: 84.7, 1.87 and 45.6 mg kg^{-1}, respectively). Mostly, all the crop growth characteristics were improved by the application of chemical fertilisers, although there was a variation between the maize landraces (Fig. 1). Application of fertiliser significantly increased the plant height by 12%, weight of cobs per plant by 26%, cob length by 23%, cob diameter by 9.6%, and grain yield by 28% over the control respectively (Fig. 1). The local germplasm usually yield lower compared to high yielding varieties [20]. This study further confirms the importance of nutrient management for young jhum fallows where the inherent soil fertility is low. Furthermore, chemical fertilisers may be dibbled near the root zone instead of broadcasting in order to avoid runoff in steep landscapes.

2.1 Conversion to Agroforestry Plantations: Effects on Soil Fertility and Health

While some regard the shifting cultivation system as an important part of a customary lifestyle for tribal groups where no inputs are required and without modern chemicals, others believe it is highly destructive and uneconomical. However, with the increasing human population and the demand for increasing food, there is an urgent need for farmers to produce more food and earn greater income from a given piece of land. The management of shifting cultivation remains a fundamental objective for agricultural development by the state governments in North-east India. Many efforts were made in the past to change the system and many alternatives were introduced to improve the system for the benefits of the farmers and the environment. Multipurpose agroforestry plantation systems have been considered as an alternative to shifting cultivation. Effect on soil properties (0–60 cm) by different agroforestry models were evaluated by Majumdar et al. [21]. It was observed that changes in soil properties were mainly attributed to soil and water conservation measures and systematic cropping patterns. The silvi-horticulture (alder (*Alnus nepalensis*; Betulaceae) + pineapple (*Ananas comosus*; Bromeliaceae + fodder grass) had the highest SOC (2.19%), followed by agri-horticulture (Assam lemon (*Citrus* × *limon*; Rutaceae) + crops; 2.02%), agri-silviculture (1.62%) and agri-horticulture (Khasi mandarin (*Citrus* × *reticulata*; Rutaceae) + crops; 1.55%) respectively.

There have been several regulations in the past to shift traditional jhum to more economical land-uses in North-east India. For example, the Mizoram State Government initiated the Village Council Act of 1953, Lushai Hills District (Jhumming) Regulation Act of 1954, Garden Colony project of 1977–78, an initial New Land Use Policy (NLUP) at a small scale in 1984–87 and a much wider scale in 1991–97 [22, 23]. With modifications from the past, the NLUP was, again, officially implemented

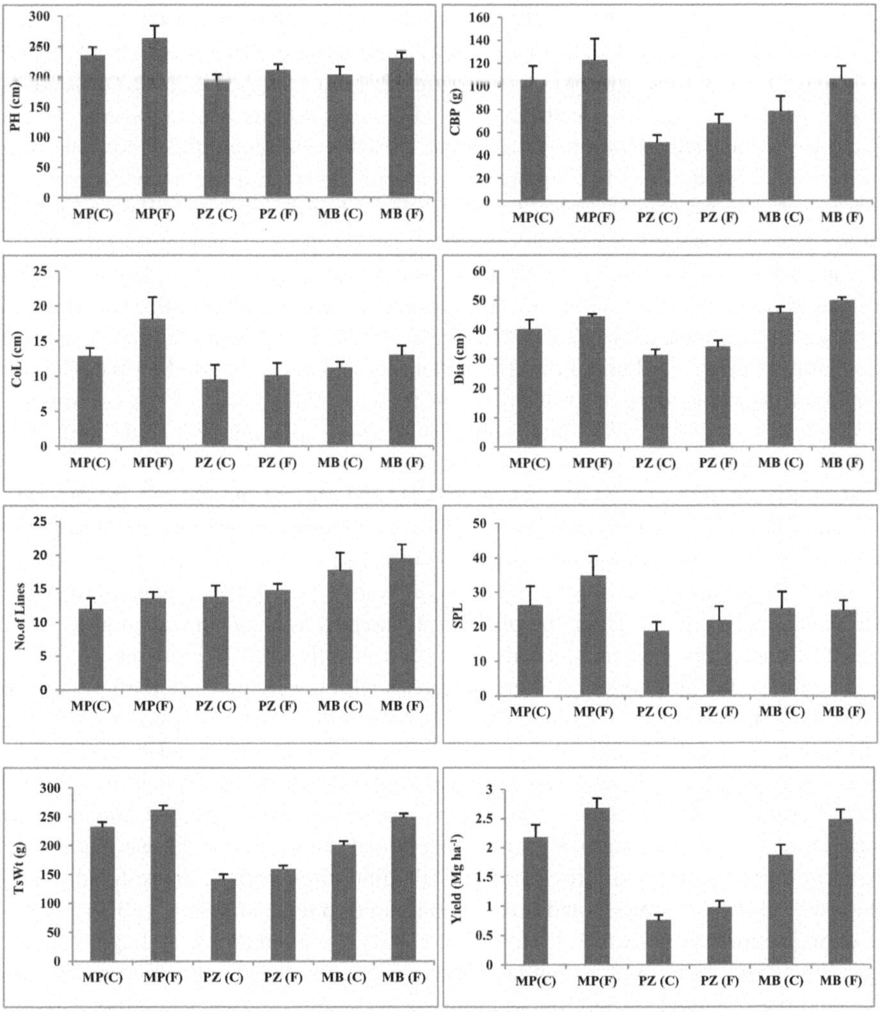

Fig. 1 Characteristics of three local maize germplasm MP (mimpui), PZ (puakzo) and MB (mimban). (C: Control; F: Chemical fertiliser; PH: Plant height; CBP: Cob weight per plant; CoL: Cob length; Dia: Cob diameter; SPL: No of seeds per line; TsWt: Test weigh, and Yield: Grain yield). Values are mean with error bars indicating standard deviations. *Source* Lungmuana et al. [19]

in 2011 in the state with an objective to abolish jhum cultivation and improve agricultural sustainability by developing alternative permanent land-use systems. The policy greatly reduced the jhum area and increased agricultural production in the state. It has been estimated that there was a reduction of jhum area and jhumias from 28,562 ha and 68,433 persons during 2010–11 to 19,851 ha and 48,417 persons during 2015–16, i.e. 30.5% reduction in area and 29.3% reduction in the number of jhumias. Due to this policy intervention, the utilisation of potential wetland rice cultivation area increased by 23.1% thereby increasing the rice productivity of the state by 31.5% during the five-year period [22]. The NLUP came to an end in 2018 with the change of government in Mizoram and has been replaced by the Socio-Economic Development Programme (SEDP). Subsequently, there has been an increase in plantations like rubber (*Hevea brasiliensis*; Euphorbiaceae), arecanut (*Areca catechu*; Arecaceae), teak (*Tectona grandis*; Lamiaceae), and oil palm (*Elaeis guineensis*; Arecaceae). Similarly, 'the slash-and-burn to sustainability project' in Nagaland improved farmer's livelihoods through horti-agroforestry plantations and soil and water conservation [24]. In Tripura state, rehabilitation of rubber plantations for shifting cultivators started as early as the 1980s, however, this initiative gained momentum during 1992 under the Block Planting Scheme supported by World Bank which led to tremendous improvements in the economic status of the community [25]. It has also been reported that rubber plantations replacing degraded jhum lands of Tripura improved soil physical, chemical and biological health [26].

Recently, there has been a great increase in the area of plantations in Mizoram and other North-eastern states of India. These plantations are established during the first-year of jhum cultivation where annual crops like rice, maize, sesame (*Sesamum indicum*; Pedaliaceae) etc. continued up to the fourth year until the planted trees are sufficiently tall (Fig. 2). Thus, the plantation also acts as a barrier to soil and runoff erosion under slopes where normal cultivation is difficult. Lungmuana et al. [27] studied the soil health under different land-uses which were converted after shifting cultivation in sub-tropical Kolasib district, Mizoram. After experiencing shifting cultivation, the land was converted to different plantations or secondary forest. The results revealed that soil physico-chemical properties were not much affected by the different land-uses, but soil biological properties were affected and used as early indicators of soil health. Recently abandoned jhum soil had a lower geometric mean of soil enzyme activities (acid and alkaline phosphatase, arylsulphatase, β-glucosidase, dehydrogenase and urease) compared to teak, arecanut, rubber plantations and secondary forest [27]. However, among the plantations, teak soil had the least SOC. The study implies that land-use change after jhum cultivation to alternative land-uses (plantation and forest) improves soil microbiological and biochemical health which was due to organic matter accumulation.

2.2 Conversion to Agroforestry Plantations: Effects on Soil Degradation and Carbon Sequestration

Soil loss under jhum cultivation system in North-east India was estimated between 170 and 200 Mg ha^{-1} year^{-1} [28]. Singh et al. [29] reported SOC loss to the tune of 6.0×10^6 Mg year^{-1} from the whole North-east region along with associated nutrient losses. In this region, degraded land was reported to be 36.6% of the total geographical area, compared to the national average of 20.2% [14, 30]. In some states like Manipur,

Fig. 2 Plantation of arecanut (*Areca catechu*; Arecaceae) during jhum cultivation at Thingdawl village, Kolasib district, Mizoram, India

Nagaland and Sikkim, more than 50% of the total geographical area comes under the degraded land category. Watershed developmental projects along with agroforestry plantations through contour farming and permanent terrace cultivation systems under jhum rehabilitation schemes play an important role in the stabilisation of runoff from soil. Under these schemes, more and more jhumia families were encouraged by providing free seedlings to the selected families through Government (both state and central Government) sponsored projects. Agroforestry can be effectively practiced on lands subjected to shifting cultivation with the aim to minimise soil erosion, tap nutrients and fix atmospheric nitrogen through leguminous tree species. These alternate cultivation practices like contour farming can be practiced with minimum permissible soil loss i.e. 0.4–15 Mg ha^{-1} year^{-1} without topsoil loss depending on the type of grass species used along the contour [31] Sharma et al. [32] studied soil loss in Sikkim under different land-use systems and revealed that barren land and agricultural fields (maize crop in terrace) had a soil loss of 166 kg ha^{-1} year^{-1} and 122 kg ha^{-1} year^{-1}, respectively, whereas cardamom (*Amomum subulatum*; Zingiberaceae) plantations and natural forest had soil losses of 66 kg ha^{-1} year^{-1} and 12 kg ha^{-1} year^{-1}.

Abandoned jhum fallows are known to have high potential for soil C capture. Ledo et al. [33] reported an average increase of 20% SOC from an annual to a perennial crop after 20 years or more in the upper soil layers. In the top 1 m of the profile, the SOC gain was 11% (±8.5%) a total of 5.66 Mg ha^{-1} in SOC. Many existing forestry-based systems are available in North-east India to assist in C capture. One among them is the alder-based system in Nagaland where the bacteria in the root nodules of the plant fix atmospheric N and increase the soil fertility. A study by Manpoong et al. [34] in Mizoram reported the enrichment of C and nutrients in the root zone thorough *Tephrosia candida* (Fabaceae) grown as hedgerow plants by increasing root exudation, litter addition and microbial proliferation. Soil C sequestration through plantations is an important strategy to limit carbon losses after conversion from shifting cultivation. Rahman et al. [35] reported that SOC sequestration can be achieved by best management practices through residue incorporation to the soil as well as increasing yield up to 33% in oil palm plantations. Conversion of forest to plantations such as tea (*Camellia sinensis*; Theaceae) and cashew (*Anacardium occidentale*; Anacardiaceae) can also increase soil C storage at an average of 0.23–0.35 Mg C ha^{-1} $year^{-1}$ [36]. A study by Ramesh et al. [37] showed that common agroforestry plantations in Meghalaya increased the soil C stock compared to no trees. Among the plantations, *Alnus nepalensis* conserved the highest C stock up to 75 cm depth (60.2 Mg ha^{-1}) followed by *Michelia oblongata* (Magnoliaceae) (54.8 Mg ha^{-1}), *Parkia timoriana* (Fabaceae) (54.3 Mg ha^{-1}) and *Pinus kesiya* (Pinaceae) (54.4 Mg ha^{-1}), compared to no trees (41.0 Mg ha^{-1}).

The proportion of active C compared to passive pools of C is affected by the land-use and the turnover of soil C depends on the management to sequester more C [38]. Due to high rainfall in the North-east region, rapid recovery of SOC after cultivation has been observed due to high biomass production and return of C via litterfall to the soil [6]. Up to 60% or even more SOC is within the active pool suggesting that this major pool of C is vulnerable to loss. Natural forest conserved more stable pools of C due to minimal disturbance and, similarly, agroforestry stored greater soil C than other agricultural systems [38].

3 Integrated Farming Systems (IFS): an Alternative to Jhum Cultivation

A sustainable system to ensure food security is required to augment the traditional systems of jhum cultivation keeping in view the economic interests of the farmers. Many studies [21, 39, 40] have suggested that mixed farming or integration of different agricultural components is beneficial for hill regions rather than concentrating on one system or component. An Integrated Farming System (IFS), as the name suggests, is a mixed farming system including selected suitable crops, agroforestry, water harvesting, animal components etc. that recycles the products for soil fertility enhancement. Different IFS models are becoming increasingly popular with the inclusion of suitable trees having a good root binding ability and producing a high quantity of litter to reduce soil water loss. Firstly, as the system aims for doubling income of the farmers, water harvesting (jalkund) of rain (Fig. 3a) and perennial streams from the hills is necessary to meet the irrigation requirement during the dry season. To maintain soil health and nutritional requirements, animal components (e.g. pigsty; Fig. 3b) are a must for manure fertiliser and fish may be reared in the pond or jalkund. Crops are selected and planted in an appropriate pattern to increase the cropping intensity and conserve soil health along with permanent silviculture and agroforestry trees for the long term. A study under the Farming System Research Project at ICAR Research Complex, Barapani, Meghalaya, on different IFSs developed over 17 years showed increases in soil fertility [39]. The study revealed that incorporation of different systems like agriculture, horticulture

Fig. 3 A jhumia with **a** water harvesting structure (jalkund) and **b** piglets in a pigsty at Thingdawl Village, Kolasib district, Mizoram, India

and silvicultural-pastoral along with soil conservation measures and judicious use of chemical fertilisers and manures increased soil fertility compared to shifting cultivation. In addition to the soil fertility increase, IFSs can reduce soil erosion up to 34–48% and organic matter loss by 26–51% compared to the traditional systems [28].

A recent study by Ray et al. [40] in Nagaland showed that water harvesting and IFS increased cropping intensity, i.e. the cultivation of more than one crop in the same field in a year, from 100% in traditional jhum to 168% due to IFS intervention and the contribution of livestock was calculated to be 57% of the net income. The IFS system increased employment opportunities seven-fold compared to the traditional system in addition to minimising soil and environmental degradation risks. The IFS component in this study included multipurpose trees like *Parkia timoriana* and alder planted on the hilltop followed by fruit trees such as assam lemon and other citrus fruits in the middle hills and bananas (*Musa* sp.; Musaceae) at the bottom. A pigsty was constructed near the jalkund to feed the fish, and poultry was reared for meat and egg purposes. Crops such as cassava (*Manihot esculena*; Euphorbiaceae), ginger, groundnut, maize, rice, rice bean (*Vigna umbellata*; Fabaceae), soybean, sweet potato (*Ipomoea batatas;* Convolvulaceae), etc. were grown as an inter-crop in between the trees in rows across the slopes. Crop management practices such as dibbling of seeds, spacing maintenance, contour bunding, half-moon terraces, biofertilisers and vermicompost etc., were used. The soil fertility improved under the IFS after two years compared to the traditional jhum cultivation which was due to soil and water conservation and recycling of crop residues thus increasing the soil organic matter and soil health.

4 Conclusions

The phases and operations involved in shifting cultivation are complex and different from typical crop cultivation. Shifting cultivation systems are still followed by many tribal groups encompassing a key cultural as well as agronomic role in their lives. However, it has been discussed that the traditional jhum cultivation is slowly shifting towards settled cultivation and the success of moving from traditional to settled cultivation can only be achieved when natural resources are conserved and farmers' livelihoods are improved. This review has, accordingly, shown how agronomic interventions can enhance soil health as well as farmers' livelihoods in North-east India. Agroforestry interventions can play a key role in enhancing the sustainability of the system compared to the traditional jhum cultivation. Integrating different farming components can increase the system productivity, reduce soil erosion and increase crop productivity. Greater awareness and further studies are required for the adoption and alteration from traditional jhum cultivation to improved settled cultivation to conserve natural resources, soil fertility and, most importantly, to improve the farmers' livelihoods. Future studies should also focus on the role of indigenous wild legumes, which are abundantly found in the region, on soil fertility, and on soil carbon sequestration potential.

References

1. Mazoyer M, Roudart L (2006) A history of world agriculture from neolithic age to the current crisis. Earthscan Publishers, London, UK
2. Brady NC (1996) Alternatives to slash-and-burn: a global imperative. Agric Ecosys Environ 58:3–11. https://doi.org/10.1016/0167-8809(96)00650-0
3. Baruah U, Bandyopadhyay S, Reza SK (2014) Land use planning and strategic measures in North Eastern region of India. Agropedology 24:292–303
4. Borggarrd O, Gafur A, Petersen L (2003) Sustainability appraisal of shifting cultivation in the Chittagong Hill Tracts of Bangladesh. Ambio 32:118–123. https://doi.org/10.1579/0044-7447-32.2.118
5. Sarkar D, Meitei CB, Baishya LK, Das A, Ghosh S, Chongloi KL, Rajkhowa DJ (2015) Potential of fallow chronosequence in shifting cultivation to conserve soil organic carbon in northeast India. Catena 135:321–327. https://doi.org/10.1016/j.catena.2015.08.012
6. Lungmuana, Singh SB, Vanthawmlian, Saha S, Dutta SK, Rambuatsaiha, Singh AR, Boopathi T (2017) Impact of secondary forest fallow period on soil microbial biomass carbon and enzyme activity dynamics under shifting cultivation in North Eastern Hill region, India. CATENA 156:10–17. https://doi.org/10.1016/j.catena.2017.03.017

7. Grogan P, Lalnunmawia F, Tripathi SK (2012) Shifting cultivation in steeply sloped regions: a review of management options and research priorities for Mizoram state, Northeast India. Agrofor Syst 84:163–177. https://doi.org/10.1007/s10457-011-9469-1
8. Choudhary VK, Singh SL, Dixit A, Arunachalam A, Bhagawati R (2014) Biomass and carbon sequestration of agroforestry trees in Arunachal Pradesh, North East India. Clim Change Environ Sustain 2(1):48–54. https://doi.org/10.5958/j.2320-642X.2.1.005
9. Devi NB, Lepcha NT, Bhutia PT, Rocky P, Sahoo UK, Pandey R, Jyoti Nath A (2023) Biodiversity and ecosystems services of the agroforestry systems of the Himalayan region: an overview. In: Dagar JC, Gupta SR, Sileshi GW (eds) Agroforestry for sustainable intensification of agriculture in Asia and Africa. Sustainability sciences in Asia and Africa. Springer, Singapore. https://doi.org/10.1007/978-981-19-4602-8_15
10. Eaton JM, Lawrence D (2009) Loss of carbon sequestration potential after several decades of shifting cultivation in the Southern Yucatan. For Ecol Manage 258:949–958. https://doi.org/10.1016/j.foreco.2008.10.019
11. Mukul SA, Herbohn J (2016) The impacts of shifting cultivation on secondary forests dynamics in tropics: a synthesis of the key findings and spatio temporal distribution of research. Environ Sci Policy 55:167–177. https://doi.org/10.1016/j.envsci.2015.10.005
12. Nye PH, Greenland DJ (1960) The soil under shifting cultivation. Technical communication No. 51. Commonwealth Bureau of Soils. Harpenden, Hertfordshire, UK
13. Shao X, Yang W, Wu M (2015) Seasonal dynamics of soil labile organic carbon and enzyme activities in relation to vegetation types in Hangzhou Bay tidal flat wetland. PLoS One 10:e0142677. https://doi.org/10.1371/journal.pone.0142677
14. Borthakur DN (1992) Agriculture of the North Eastern region with special reference to hill agriculture. Beecee Prakashan, Guwahati, Assam, India
15. Tripathi SK, Vanlalfakawma DC, Lalnunmawia F (2017) Shifting cultivation on steep slopes of Mizoram, India: impact of policy reforms. In: Cairns M (ed) Shifting cultivation policies: balancing environmental and social sustainability. CABI, Wallingford, UK, pp 393–413. https://doi.org/10.1079/9781786391797.0393
16. Wapongnungsang, Ovung, EY, Upadhyay KK, Tripathi SK (2021) Soil fertility and rice productivity in shifting cultivation: impact of fallow lengths and soil amendments in Lengpui, Mizoram northeast India. Heliyon 7:e06834. https://doi.org/10.1016/j.heliyon.2021.e06834
17. Baishya LK, Ansari MA, Sarkar D, Ghosh M, Kumar S, Prakash N (2016) Productivity enhancement in shifting cultivated lands through biofertilizer and micro-dosing of NPK in Eastern Himalayan Region. Res Crop 17:268–275. https://doi.org/10.5958/2348-7542.2016.00046.2
18. Choudhary VK, Kumar PS, Kanwat M, Bhagawati R (2012) Improvement of jhum with crop model and carbon sequestration techniques to mitigate climate change in Eastern Himalayan region, India. J Agric Sci 4:181–189. https://doi.org/10.5539/jas.v4n4p181
19. Lungmuana, Singh AR, Ramakhrishna Y, Rambuatsaiha, Saha S, Dutta SK, Boopathi T, Vanthawmliana, Shakuntala I, Dayal V, Singh SB (2019) Evaluation of nutrient management on three major maize landrace characteristics of Mizoram. In: Saha S, Lungmuana, Dayal V, Ansari MA, Jat SL, Chowdhury S, Lalliansanga S, Shakuntala I, Saithantluanga H, Rakshit S, Prakash N (eds) Souvenir: national workshop on scientific maize cultivation in North East India, March 05 2019, Aizawl, Mizoram. ICAR Research Complex for NEH Region, Umiam, Meghalaya, India, pp 44–48
20. Sibiya J, Tongoona P, Derera J, Makanda I (2013) Farmers‘ desired traits and selection criteria for maize varieties and their implications for maize breeding: a case study from KwaZulu-Natal Province, South Africa. J Agric Rural Dev Trop Subtrop 114:39–49
21. Majumdar B, Kumar K, Venkatesh MS, Patiram BBP (2004) Effect of different agroforestry systems on soil properties in acid Alfisols of Meghalaya. J Hill Res 17:1–5
22. Zothansanga D, Beingachhi B (2019) The New Land Use Policy: a panacea for shifting cultivation in Mizoram. Senhri J Multidiscip Stud 4(1):21–29. https://doi.org/10.36110/sjms.2019.04.01.003
23. Thanga JLT (2020) Land use policies in the state of Mizoram. J Econ Soc Dev 16:55–95

24. Panda BK, Alam MA, Sarkar S (2016) Shifting cultivation in North-East India: trend, benefit and challenges. IUT J Adv Res Dev 2:85–114
25. Roy M, Saha S, Das J, Roy M (2015) Transition from shifting cultivation to rubber cultivation in Tripura: socio-economic and ecological impact. Clarion 4:64–74
26. Krishnakumar AK, Potty SN (1992) Nutrition of *Hevea*. In: Sethuraj MR, Mathew NM (eds) Natural rubber: biology, cultivation and technology, Elsevier, Amsterdam, The Netherlands, pp 239–262
27. Lungmuana, Singh SB, Choudhury BU, Vanthawmliana, Hnamte V (2019) Transforming jhum to plantations: effect on soil microbiological and biochemical properties in the foot hills of North Eastern Himalayas, India. Catena 177:84–91. https://doi.org/10.1016/j.catena.2019.02.008
28. Chatterjee D, Kuotsu R, Ray SK, Patra MK, Thirugnanavel A, Kumar R, Borah TR, Chowdhury P, Pongen I, Satapathy BS, Deka BC (2022) Preventing soil degradation in shifting cultivation using integrated farming system models. Arch Agron Soil Sci 68:1841–1857. https://doi.org/10.1080/03650340.2021.1937139
29. Singh NP, Singh OP, Jamir NS (1996) Sustainable agriculture development strategy for North Eastern hill region of India. In: Shukla SP, Sharma N (eds) Sustainable development strategy. Mittal Publications, New Delhi, India, pp 346–351
30. Anonymous (2000) Wastelands Atlas of India. Ministry of Rural Development, Government of India, New Delhi, India and National Remote Sensing Agency, Hyderabad, India
31. Thansanga R (1997) New contour farming system. Ind J Soil Conserv 25:167–169
32. Sharma E, Sundriyal RC, Rai SC, Bhatt YK, Rai LK, Sharma R, Rai YK (1992) Integrated watershed management: a case study in Sikkim Himalaya. Gyanodaya Prakashan, Nainital, Uttar Pradesh, India
33. Ledo A, Smith P, Zerihun A, Whitaker J, Vicente-Vicente JL, Qin Z, McNamara NP, Zinn Y, Llorente M, Liebig M, Kuhnert M, Dondini M, Don A, Diaz-Pines E, Datta A, Bakka H, Aguilera E, Hillier J (2020) Changes in soil organic carbon under perennial crops. Glob Change Biol 26:4158–4168. https://doi.org/10.1111/gcb.15120
34. Manpoong C, Hauchhum R, Tripathi SK (2020) Soil fertility and root carbon exudation in *Tephrosia candida* (Roxb.) DC hedgerows under Sloping Agricultural Land Technology in Mizoram, northeast India. J Trop Agric 58:12–21
35. Rahman N, Giller KE, de Neergaard A, Magid J, van de Ven G, Bruun TB (2021) The effects of management practices on soil organic carbon stocks of oil palm plantations in Sumatra, Indonesia. J Environ Manage 278:111446. https://doi.org/10.1016/j.jenvman.2020.111446
36. Mishra G, Sarkar A, Giri K, Nath AJ, Lal R, Francaviglia R (2021) Changes in soil carbon stocks under plantation systems and natural forests in Northeast India. Ecol Model 446:109400. https://doi.org/10.1016/j.ecolmodel.2021.109500
37. Ramesh T, Manjaiah KM, Mohopatra KP, Rajasekar K, Ngachan SV (2015) Assessment of soil organic carbon stocks and fractions under different agroforestry systems in subtropical hill agroecosystems of north-east India. Agrofor Syst 89:677–690. https://doi.org/10.1007/s10457-015-9804-z
38. Sahoo UK, Singh SL, Gogoi AA, Kenye A, Sahoo SS (2016) Active and passive soil organic carbon pools as affected by different land use types in Mizoram, Northeast India. PLoS One 14:e0219969. https://doi.org/10.1371/journal.pone.0219969
39. Majumdar B, Venkatesh MS, Satapathy KK, Kumar K, Patiram (2002) Effect of alternative farming systems to shifting cultivation on soil fertility. Indian J Agri Sci 72:122–124
40. Ray SK, Chatterjee D, Rajkhowa DJ, Baishya SK, Hazarika S, Paul S (2020) Effects of integrated farming system and rain water harvesting on livelihood improvement in North-Eastern region of India compared to traditional shifting cultivation: evidence from an action research. Agrofor Syst 94:451–464. https://doi.org/10.1007/s10457-019-00406-3

Changes in Bacterial Community Structure and Nitrogen Cycling Gene Abundance During the Fallow Recovery Phase of Shifting Cultivation in North-East India

Francis Q. Brearley, Max W. Spencer, Binu M. Tripathi, Etsoshan Y. Ovung, Ng. Somen Singh, Krista L. McGuire, and Shri Kant Tripathi

Abstract We have a good understanding of the recovery of vegetation during secondary succession of the fallow phase of shifting cultivation, but much less knowledge on how this process influences important soil micro-organisms. This study examined changes in the bacterial community (Illumina sequencing) and gene function (qPCR) in fallows of 1–15 years of age in Mizoram, North-east India, and compared them to a natural forest. We found that there was little change in the bacterial community during the first 10 years of fallow recovery but that bacteria in fallows of 15 years of age were more similar to natural forests. In terms of individual phyla, Cyanobacteria were most abundant at the early stages of succession whilst Proteobacteria were positively correlated, and Firmicutes were negatively correlated with soil carbon. There was no change in the abundance of *amoA* and *nosZ* genes during succession. Our results indicate that 15 years of fallow regrowth may be sufficient for bacterial communities to recover their community structure as long as this is accompanied by increases in soil carbon and nutrients.

F. Q. Brearley (✉)
Department of Natural Sciences, Manchester Metropolitan University, Chester Street, Manchester M1 5GD, UK
e-mail: f.q.brearley@mmu.ac.uk

M. W. Spencer · K. L. McGuire
Department of Biology, Institute of Ecology and Evolution, University of Oregon, Eugene, OR, USA
e-mail: mws@oregon.edu

K. L. McGuire
e-mail: kmcguire@uoregon.edu

B. M. Tripathi
Korea Polar Research Institute, Incheon 21990, Republic of Korea

E. Y. Ovung · Ng. S. Singh · S. K. Tripathi
Department of Forestry, Mizoram University, Aizawl, Mizoram 796004, India

B. M. Tripathi
Division of Plant and Soil Sciences, West Virginia University, Morgantown, WV 26506, USA

S. K. Tripathi and F. Q. Brearley (eds.), *Shifting Cultivation Systems*,
https://doi.org/10.1007/978-3-031-70388-1_12

Keywords Functional genes · Microbial communities · Secondary succession · Soils

1 Introduction

Slash-and-burn (aka swidden or jhum) agriculture plays a fundamental role in the lifestyle and traditions of many communities around the world including tribal communities in North-east India [1]. During slash-and-burn, forest is cut and burned with many of the nutrients contained within the ash fertilising the subsequent crops. In most cases, the field is cultivated for a few years then left fallow to recover its nutrient stocks, with the recovery relating to the growth rate of the fallow forest during secondary succession. We have a good understanding of the rate of recovery of forest above- and below-ground biomass as well as the plant/tree community structure during the process of secondary succession in North-east India [2–4]. We also have a reasonable number of studies on recovery of soil chemistry and nutrient stocks [3–5]. However, changes in soil biological parameters during fallow forest recovery are much less studied and, in North-east India, have received scant attention. To understand the processes of ecosystem recovery following disturbance, knowing how below-ground communities change is important due to their roles in key biogeochemical pathways and processes [6]. In addition to examining community composition and structure, it is also important to determine how the below-ground community functions, and one way that we can do this when examining microbial communities is through analysing the abundance of particular genes [7]. In assessing the abundance of functional genes, we can elucidate the potential of the community to perform a function of interest [7, 8]. Within below-ground communities, functions of interest often include the cycling of nutrients such as nitrogen due to its clear effect on ecosystem productivity.

Soil bacteria are one such incredibly diverse group of taxa in the soil environment but they are clearly sensitive to changes in land-use [9]. These land-use changes induce alterations to vegetation composition as well as to soil physical and chemical properties that may be mediated by land management practices such as fertilisation. We know that bacteria are influenced by a multitude of factors including soil properties, vegetation composition, climate and soil depth [6]. Furthermore, bacterial communities are often considered to be more dynamic than changes to soil physico-chemical properties and therefore may be suitable as indicators of changes in below-ground ecological function [10]. Whilst 'macroscopic' taxa usually show a decrease in diversity following forest loss or conversion [11], the changes to soil-dwelling taxa is less well understood. Nevertheless, meta-analyses have shown that soil bacteria regularly increase in diversity follow land-use change [9], that particular taxa consistently decline in abundance [9, 12] and that the communities often show a homogenisation in their structure [13]. Furthermore, bacterial communities appear to be more influenced by land-use change than fungi [14, 15]. Whilst these studies were from tropical environments, they have examined a range of land-use changes but rarely shifting cultivation and the subsequent succession to secondary forest. In this chapter, we were interested to examine changes in soil bacterial communities during the secondary succession fallow recovery phase in North-east India and explore aspects of their functioning.

2 Methods

2.1 Study Site

Soils were collected from around Reiek Community Forest and the adjacent Ailawng village in Mizoram, North-east India (23° 41′ N, 92° 41′ E) that experiences a wet subtropical climate with a precipitation of 2680 mm and temperatures ranging between 8–22 °C in winter and 20–28 °C in summer [5]. In this location, common crops include upland paddy (*Oryza sativa*; Poaceae), maize (*Zea mays*; Poaceae), tomatoes (*Solanum lycopersicum*; Solanaceae), *Capsicum* species (Solanaceae), taro (*Colocasia esculenta*; Araceae), bananas (*Musa* species; Musaceae), ginger (*Zingiber officinale*; Zingiberaceae) and turmeric (*Curcuma longa*; Zingiberaceae) and the area is normally farmed for one to two years before being allowed to recover naturally.

2.2 Sampling

Samples were taken from seventeen locations in November 2018 (800–1200 m asl) There were three sites 1-year-old, two sites 2-years-old, three sites 5-years-old, three sites 8–10-years-old, three sites 15-years-old, and three sites from a natural (primary) forest (Fig. 1). Ages were determined by personal observations of recent farming activities and in consultation with a knowledgeable local community member for older fallows. From each location, soils (0–5 cm depth) were taken from the four corners of a 1 m^2 quadrat and composited for further analysis. Fresh soils were placed in Lifeguard preservation solution before DNA was extracted from 0.25 g of each sample using a MoBio PowerSoil kit following the manufacturer's instructions with 1 min in a FastPrep during the first step [16].

2.3 DNA Sequencing

The V4 region of 16S rRNA gene was amplified using the primer pair 515F-806R [17] with the relevant barcodes. All PCR reactions were carried out with Phusion® High-Fidelity PCR Master Mix (New England Biolabs). PCR products were mixed in an equimolar ratio and then purified with Qiagen Gel Extraction Kit (Qiagen,

Fig. 1 Secondary forest fallows of **a** one-year and **b** five-years old and a **c** natural forest sampled for their soil bacterial communities in Mizoram, North-east India

Germany). Libraries were generated with NEBNext® UltraTM DNA Library Prep Kit for Illumina and quantified via Qubit and qPCR, before analysis on an Illumina platform with 250-bp paired-end chemistry. Sequencing was conducted by Novogene (China).

2.4 Quantitative PCR (qPCR) of Nitrogen-Cycling Genes

qPCR was used to determine the concentration of two genes involved in nitrogen cycling within each sample, *amoA* (ammonia monooxygenase) and *nosZ* (nitrous oxide reductase). Previously created primers were used for the genes of interest: *amoA* (amoA-1F-GC [18] and amoA-2R [19]) and *nosZ* (nosZF and nosZ1622R [20]). To generate standard curves, double stranded DNA constructs known as gBlocks, obtained from Integrated DNA Technologies (Idaho, USA), were used. qPCR was performed on a BioRad CFX96 real-time PCR instrument (BioRad, California, USA); each reaction contained 5 μL SYBR Green Mastermix, 250 nM forward primer, 250 nM reverse primer, 1 ng of sample template DNA, and enough H_2O for a total reaction volume of 10 μl. qPCR conditions were as follows: initial denaturation at 94 ºC for 2 min followed by a 40-cycle two-step reaction of 94 ºC for 15 s, followed by 60 ºC for 1 min. All reactions were performed in triplicate.

2.5 Soil Analyses

The moisture content of fresh soil was determined by heating subsamples to 105 °C for 24 h and the remainder was air-dried and ground to pass a 1-mm sieve for chemical analyses. Soil pH was measured by adding 5 g of fresh soil to 12.5 ml of deionised water; the mixture was then shaken and left to equilibrate for 24 h before measurement with a Mettler-Toledo pH meter. Total carbon and nitrogen were determined on a Vario EL Cube elemental analyser. Phosphorus was extracted from

2.5 g sub-samples that were shaken with 25 ml of Mehlich 3 solution for ten minutes before being filtered and analysed on a Thermo iCAP 6300 Duo inductively coupled plasma optical emission spectrometer (ICP-OES) with correction by determining the moisture content of the air-dried soil that were heated as above. Cations were leached from 5 g soil samples with 100 ml of ammonium acetate, diluted ten-fold and analysed on an ICP-OES as above.

2.6 Bioinformatics

The paired-end 16S rRNA gene sequences were analysed in Mothur v.1.40.5 [21] using an approach very similar to that of Tripathi et al. [22]. First, sequences were assembled into contigs, and sequences with read length than <150 bp and base-call ambiguity were removed. Sequences were then aligned against a SILVA alignment (https://www.mothur.org/wiki/Silva_reference_files) and pre-clustered using a pseudo-single linkage algorithm [23]. Chimeric sequences were removed using VSEARCH in de novo mode [24]. Taxonomic classifications were performed using a naive Bayesian classifier (80% bootstrap cutoff with 100 iterations) against the EzTaxon-extended database [25]. Sequences were clustered into operational taxonomic units (OTUs) at a threshold of $\geq$ 97% similarity using the OptiClust algorithm [26]. A mean of 132,920 $\pm$ (s.d.) 34,339 sequences were found per sample. All singleton OTUs were removed to avoid inclusion of sequences likely due to sequencing errors. The sequences were deposited in NCBI's Sequence Read Archive under the BioProject accession number: PRJNA905879.

2.7 Statistics

To analyse the variations in soil physico-chemical properties, phylum abundance and alpha-diversity indices of bacterial community between soil fallow age groups, we used one-way ANOVAs (across all soil ages of secondary forest fallows and natural forest) followed by additional post hoc Tukey's tests where required or *t*-tests (between two groups: all secondary forests combined and natural forest) using Minitab 19.2020.1. The pairwise Bray–Curtis dissimilarities between communities were visualised on a non-metric multi-dimensional scaling (NMDS) plot using the vegan R package [27]. A permutational multivariate ANOVA (PerMANOVA) test was carried out to compare the Bray–Curtis dissimilarities between soil fallow age groups using PRIMER 6.0 with the PerMANOVA+ add-on [28, 29]. The association between the bacterial community and soil properties were analysed by fitting the vectors of soil properties onto NMDS ordination space using the 'envfit' function of vegan [27]. Correlations between measures used a Spearman's correlation in Minitab 19.2020.1.

3 Results

In total, 16,744 bacterial OTUs were found across all samples with a mean OTU richness of 3886 ± (s.d.) 914 per sample. Diversity metrics did not differ between fallows of different ages or natural forest with a mean Shannon diversity index of 3.10 ± (s.d.) 0.48 and a Chao1 richness of 5136 (s.d.) 980 ($p > 0.15$ in both cases).

The taxa found showed a broad representation across the bacterial tree of life with most OTUs found in Proteobacteria (29.9%), Chloroflexi (12.8%) and Actinobacteria (12.3%) (Fig. 2). In terms of numbers of sequences, the most abundant phylum was the Proteobacteria (many of which were in the Enterobacteriaceae) which made up 81% of all the sequences (Fig. 3). We saw few significant patterns of changes in phylum composition with fallow age with the exception of Cyanobacteria that were four times more abundant in the 1-year-old compared to 2-year-old fallows ($p < 0.001$) and an order of magnitude more abundant in 1-year-old fallows than all other ages, and Fusobacteria that were most abundant (two orders of magnitude) in 5- and 10-year-old fallows ($p = 0.035$) (Fig. 3). However, when samples were grouped into fallow and natural forest groups, we found that Firmicutes ($p = 0.017$) Armatimonadetes ($p = 0.011$) and Fusobacteria ($p = 0.034$) were less abundant in the natural forest compared to secondary fallow forests (Fig. 3). The NMDS ordination showed that the community structure differed by fallow age with the natural forest samples clearly distinct from the majority of the younger fallows (Fig. 4). Although the perMANOVA showed that there was no significant effect of fallow ages when they were considered separately, the bacterial community of the fallows grouped together was significantly distinct from that of the natural forest ($p = 0.003$).

Soil pH did not differ among forest age classes but soil C and N and Ca were significantly greater in the natural forest (C in 15-year-old fallow was the same as the natural forest); P and Mg were also greatest in the natural forest but not significantly so. Potassium did not clearly differ by forest age (Table 1). The envfit showed that soil C, N and Ca were significant predictors of the bacterial community composition.

There was a significant negative correlation between soil C and N and Shannon diversity ($\rho = 0.83$ and 0.68 and $p < 0.001$ and 0.006, respectively) and this relationship was similar but not as strong for OTU richness or Chao 1 diversity metrics ($p < 0.10$ in all cases). When corrected for multiple comparisons, the only phyla to show a relationship with soil properties was the abundance of Proteobacteria which was positively correlated with soil C ($\rho = 0.82$) and that of Firmicutes which was negatively correlated with soil C ($\rho = -0.75$).

There were no significant differences in nitrogen cycling gene abundances during succession (Table 2); this pattern was also seen when the gene abundance was normalised to the amount of DNA extracted (data not shown).

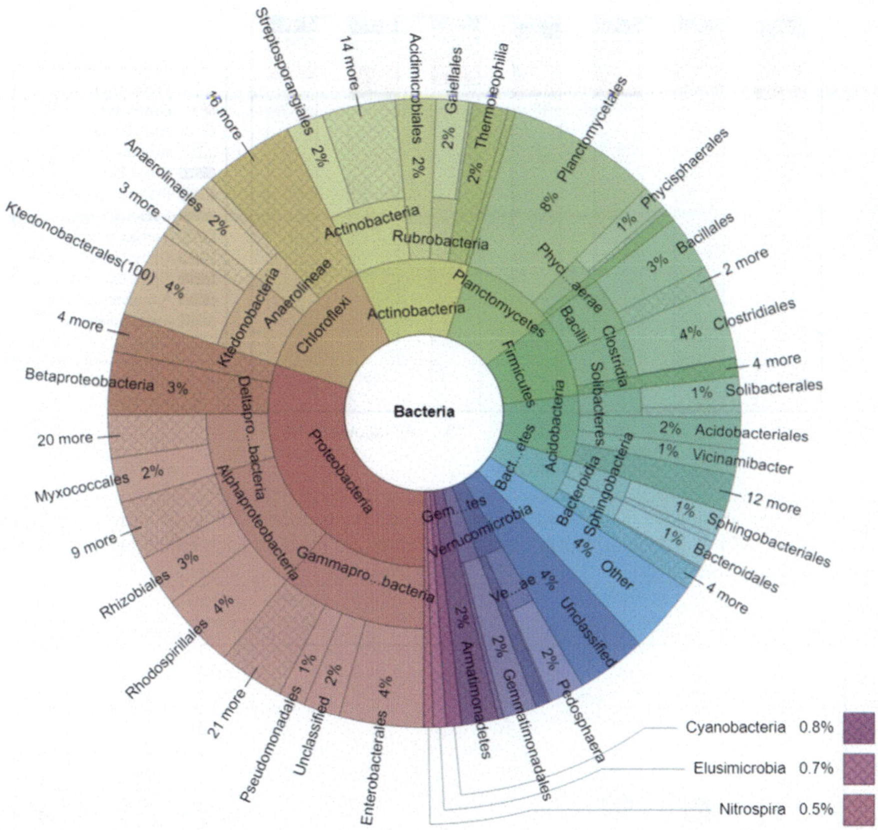

Fig. 2 Krona diagram of the abundance of soil bacterial taxa in secondary forest fallows of various ages and natural forest of Mizoram, North-east India

4 Discussion

We show here how the bacterial community changes following secondary succession of post-shifting agricultural fallows in North-east India. Although the fallow bacterial community, when all ages were combined, was significantly different from the reference natural forest in the abundance of a number of bacterial taxa, there were no clear directional changes in the bacterial community over at least the first 10 years of succession, with the exception of a reduction in cyanobacterial abundance, suggesting that stand age per se is not the key determinant of the changes in bacterial community composition at the early stages of succession. After 15-years of succession, however, the bacterial community did not differ from that in the natural forest. These results are in line with Zhang et al. [30] who found stand age only explained a small proportion (< 10%) of variation in the bacterial community in

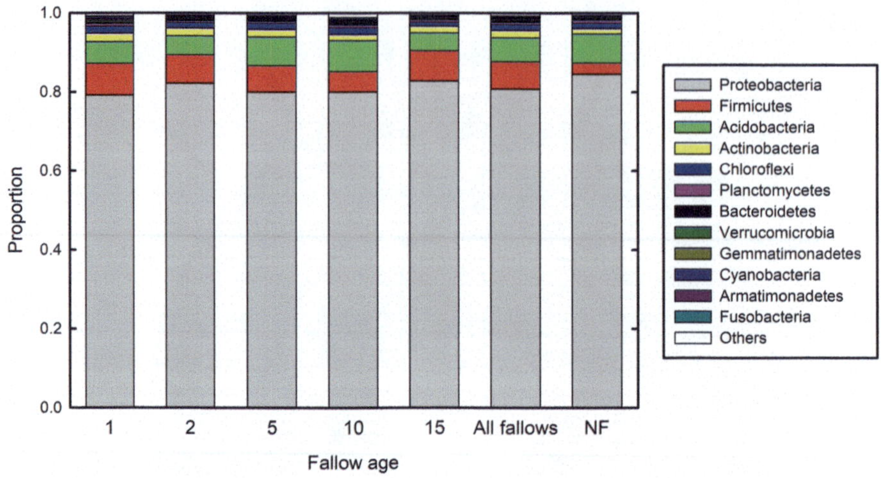

Fig. 3 Relative abundance of soil bacterial phyla in secondary forest fallows of various ages and natural forest (NF) of Mizoram, North-east India. Values are mean of each fallow age (error bars are not presented for clarity)

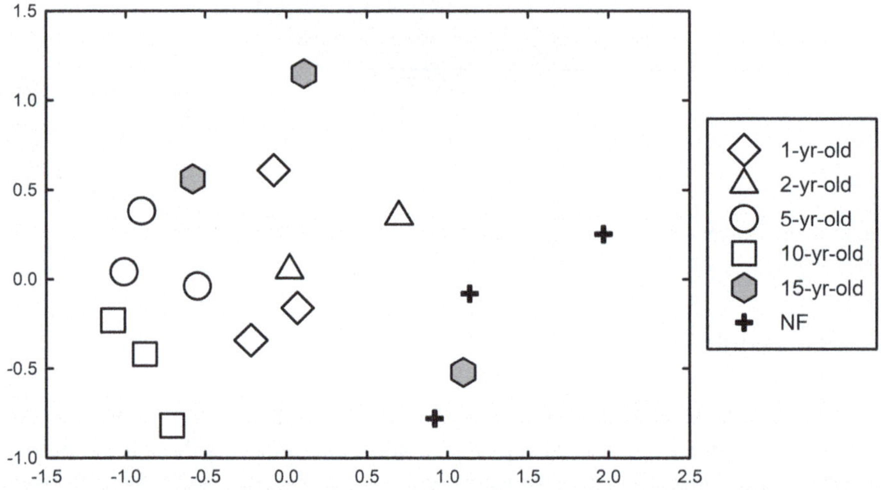

Fig. 4 NMDS ordination of soil bacterial communities in secondary forest fallows of various ages and natural forest (NF) of Mizoram, North-east India

central China. In contrast, there were clear changes in the plant community composition which increased in biomass and diversity during succession at this location [3, 4]. We also found that soil properties did not show major changes in fertility-related metrics during succession (excepting the natural forest that was, in general, most

Table 1 Soil properties in secondary forest fallows of various ages and natural forest (NF) of Mizoram, North-east India

	Age					
	1	2	5	10	15	NF
pH	5.11 ± 0.17 a	n/a	4.94 ± 0.06 a	5.05 ± 0.13 a	4.78 ± 0.21 a	5.57 ± 0.30 a
C (%)	3.30 ± 0.73 b	4.37 ± 1.44 b	3.84 ± 1.37 b	4.09 ± 0.23 b	5.34 ± 0.28 ab	8.81 ± 0.56 a
N (%)	0.27 ± 0.02 b	0.31 ± 0.05 b	0.29 ± 0.09 b	0.33 ± 0.01 b	0.42 ± 0.03 b	0.71 ± 0.06 a
P (mg kg^{-1})	2.01 ± 0.18 b	2.69 ± 044 ab	3.87 ± 0.29 ab	3.13 ± 0.82 ab	4.65 ± 0.97 ab	10.3 ± 4.12 a
K ($cmol_c$ kg^{-1})	0.41 ± 0.06 a	0.55 ± 0.08 a	0.23 ± 0.10 a	0.35 ± 0.05 a	0.30 ± 0.05 a	0.39 ± 0.06 a
Ca ($cmol_c$ kg^{-1})	2.90 ± 1.07 b	1.60 ± 0.84 b	1.79 ± 1.07 b	2.40 ± 0.58 b	2.49 ± 1.58 b	12.8 ± 2.26 a
Mg ($cmol_c$ kg^{-1})	2.00 ± 0.72 ab	1.31 ± 0.58 ab	1.04 ± 0.54 b	1.83 ± 0.59 ab	1.52 ± 0.58 ab	4.02 ± 0.37 a

Letters indicate significant differences according to Tukey's tests; n/a indicates data not available

Table 2 Abundance of *amoA* (×10^4 and *nosZ* (×10^8) genes per gram of soil in secondary forest fallows of various ages and natural forest (NF) of Mizoram, North-east India

	Age					
	1	2	5	10	15	NF
amoA	7.00 ± 1.80	2.10 ± 0.55	4.21 ± 1.58	7.97 ± 2.84	11.7 ± 1.78	3.77 ± 0.55
nosZ	1.44 ± 0.20	0.34 ± 0.03	0.34 ± 0.05	0.49 ± 0.15	1.01 ± 0.19	0.75 ± 0.04

fertile) although, again, some soil parameters (notably, soil C) were similar in the 15-year-old secondary forest compared to the natural forest.

Our work shows some similarities with other studies that were included in a meta-analysis by Petersen et al. [9]. Overall, Petersen et al.'s study showed that land-use change in the tropics led to decreases in the abundance of Acidobacteria and Proteobacteria, and increases in Actinobacteria and Chloroflexi as well as increases in Firmicutes (in contrast to our study that found significant declines). Another meta-analysis by Díaz-Vallejo et al. [12] found declines in Proteobacteria and increases in Actinobacteria as above, but they also found secondary forests had increases in Firmicutes as we found. The decline in cyanobacterial abundance during succession is related to their important role in early successional ecosystems where their photosynthetic activity contributes carbon to the soil [31, 32] and they were likely part of biocrusts in the early post shifting cultivation fields. The only comparable study in this geographical region is that of Manpoong et al. [33] who showed that Acidobacteria and Proteobacteria were most abundant in natural forest compared to other land-uses and that Verrucomicrobia were most abundant in shifting cultivation fallows. Of course, each study in earlier meta-analyses [9, 12] showed individual patterns that will depend on various factors including the type and intensity of the disturbance as Lin et al. [14], for example, showed few consistent changes during secondary succession for the most abundant bacterial phyla after shifting cultivation in New Guinea. This could be attributed to the minimal soil disturbance during low-intensity shifting cultivation in New Guinea compared to, for example, conversion to agro-industrial ecosystems [34]. We did not see an increase in bacterial alpha diversity after land-use change as commonly found in other studies [9]. However, if homogenisation of the bacterial community occurs [13], diversity at the landscape level will have been reduced.

We did not see clear changes in many soil chemical properties in contrast to earlier work in the same study location [3–5] but the variability among age classes was high—probably due to our spatially dispersed sampling scheme encompassing fields with a range of management strategies. Soil carbon and nitrogen were most important in influencing the bacterial community as might be expected given their important roles in in regulating soil microbial biomass and metabolism [35–37]. Calcium was also notably higher in the natural forest (there was no indication of carbonate rocks here as they are sandstone [38]). In particular, the abundance of Proteobacteria and Firmucutes was strongly correlated with soil carbon which can explain the patterns found in Petersen et al.'s [9] meta-analysis as land-use change regularly leads to a loss of soil carbon [39]. Whilst there was a decline in soil carbon in fallows, community changes may be due to an increase in labile carbon fractions [40, 41] which influence bacterial diversity [9] and this would be a valuable addition to further studies as it is known that different bacterial phyla have different functional responses to the addition of labile carbon such as manure [42].

It was surprising that we did not see changes in *amoA* (i.e. ammonia oxidation as a key step in nitrification) or *nosZ* (nitrous oxide reduction as a key step in denitrification) gene abundance as there were changes in soil carbon and nitrogen during succession, and an earlier study at this location showed clear increases in mineral nitrogen availability and nitrification during succession [4]. In a related study, Sun & Badgley [15] found that *amoA* was the only key nitrogen cycling gene to increase during post-mining secondary succession whilst other genes did not. One shortcoming of our approach is that gene abundance may not directly correlate to function, as there can be differences in expression between species and individuals [43]. Additionally, qPCR does not distinguish between active and dormant cells. The use of metatranscriptomics may help in understanding this disconnect between gene abundance and nitrogen cycling patterns within these sites [44]. Furthermore, there may be other genes playing a larger role in nitrogen cycling in these sites, such as *nifH* that is involved in nitrogen fixation [45].

Given sufficient time, successional processes can return the species composition and soil properties to that resembling an undisturbed ecosystem. In the bacterial community, there were, however, indications of a directional change as the community of 15-year-old secondary forests was not different from that in the natural forest which may be related to the greater soil carbon and nitrogen concentrations in these older secondary forests. This agrees with work from Brazil where Aboim et al. [46] found that the bacterial communities of older forests (15 years or older) were more similar to each other than younger forests or currently cropped areas. Therefore, ten years of succession is unlikely to be sufficient to lead to a bacterial community similar to that of natural forest but after 15 years of succession there are indications that the communities are converging with the natural forest in their composition—particularly if there is rapid accumulation of soil carbon and nutrients.

Acknowledgements We thank colleagues at Novogene for sequencing, David McKendry for assistance with soil analyses, Katherine Shek for assisting with qPCR, and QR GCRF funding from Manchester Metropolitan University that supported this work.

References

1. Cairns MF (ed) (2015) Shifting cultivation and environmental change: indigenous people, agriculture and forest conservation. Routledge, Abingdon, Oxon, UK
2. Gogoi A, Sahoo UK, Saikia H (2020) Vegetation and ecosystem carbon recovery following shifting cultivation in Mizoram-Manipur-Kachin rainforest eco-region, southern Asia. Ecol Process 9:21. https://doi.org/10.1186/s13717-020-00225-w

3. Sharma SB, Kumar S, Ovung EY, Konsam B (2022) Vegetation dynamics and soil nutrients across different shifting cultivation fallows in montane subtropical forest of Mizoram, NE India. Acta Oecol 115:103833. https://doi.org/10.1016/j.actao.2022.103833
4. Singha D, Brearley FQ, Tripathi SK (2020) Fine root and soil nitrogen dynamics during stand development following shifting agriculture in Northeast India. Forests 11:1236. https://doi.org/10.3390/f11111236
5. Ovung EY, Tripathi SK (2021) Recovery of soil carbon and nutrients in forest chronosequence following jhum cultivation in Mizoram, Northeast India. Environ Ecol 38:569–577
6. Lladó S, López-Mondéjar R, Baldrian P (2017) Forest soil bacteria: diversity, involvement in ecosystem processes and response to global change. Microbiol Mol Biol Rev 81:e00063-e116. https://doi.org/10.1128/MMBR.00063-16
7. Levy-Booth DJ, Prescott CE, Grayston SJ (2014) Microbial functional genes involved in nitrogen fixation, nitrification and denitrification in forest ecosystems. Soil Biol Biochem 75:11–25. https://doi.org/10.1016/j.soilbio.2014.03.021
8. Morales S, Cosart T, Holben W (2010) Bacterial gene abundances as indicators of greenhouse gas emission in soils. ISME J 4:799–808. https://doi.org/10.1038/ismej.2010.8
9. Petersen IAB, Meyer KM, Bohannan BJM (2019) Meta-analysis reveals consistent bacterial responses to land use change across the tropics. Front Ecol Evol 7:391. https://doi.org/10.3389/fevo.2019.00391
10. Barnard R, Osborne C, Firestone M (2013) Responses of soil bacterial and fungal communities to extreme desiccation and rewetting. ISME J 7:2229–2241. https://doi.org/10.1038/ismej.2013.104
11. Gibson L, Lee TM, Koh LP, Brook BW, Gardner TA, Barlow J, Peres CA, Bradshaw CJA, Laurance WF, Lovejoy TE, Sodhi NS (2011) Primary forests are irreplaceable for sustaining tropical biodiversity. Nature 478:378–381. https://doi.org/10.1038/nature10425
12. Díaz-Vallejo EJ, Seeley M, Smith AP, Marín-Spiotta E (2021) A meta-analysis of tropical land-use change effects on the soil microbiome: emerging patterns and knowledge gaps. Biotropica 53:738–752. https://doi.org/10.1111/btp.12931
13. Rodrigues JLM, Pellizari VH, Mueller R, Baek K, da Jesus EC, Paula FS, Mirza B, Hamaoui GS Jr, Tsai SM, Feigl B, Tiedje JM, Bohannan BJM, Nüsslein K (2013) Conversion of the Amazon rainforest to agriculture results in biotic homogenization of soil bacterial communities. Proc Natl Acad Sci USA 110:988–993. https://doi.org/10.1073/pnas.1220608110
14. Lin Q, Baldrian P, Li L, Novotny V, Heděnec P, Kukla J, Umari R, Meszárošová L, Frouz J (2021) Dynamics of soil bacterial and fungal communities during the secondary succession following swidden agriculture in lowland forests. Front Microbiol 12:676251. https://doi.org/10.3389/fmicb.2021.676251
15. Sun S, Badgley BD (2019) Changes in microbial functional genes within the soil metagenome during forest ecosystem restoration. Soil Biol Biochem 135:163–172. https://doi.org/10.1016/j.soilbio.2019.05.004
16. Brearley FQ (2020) Sequence data describing the fungal community in a tropical quartzite soil. Data Br 29:105112. https://doi.org/10.1016/j.dib.2020.105112
17. Caporaso JG, Lauber CL, Walters WA, Berg-Lyons D, Huntley J, Fierer N, Owens SM, Betley J, Fraser L, Bauer M, Gormley N, Gilbert JA, Smith G, Knight R (2012) Ultra-high-throughput microbial community analysis on the Illumina HiSeq and MiSeq platforms. ISME J 6:1621–1624. https://doi.org/10.1038/ismej.2012.8
18. Stephen JR, McCaig AE, Smith Z, Prosser JI, Embley TM (1996) Molecular diversity of soil and marine 16s rRNA gene sequences related to beta-subgroup ammonia-oxidizing bacteria. Appl Environ Microbiol 62:4147–4154. https://doi.org/10.1128/aem.62.11.4147-4154.1996
19. Rotthauwe J-H, Witzel K-P, Liesack W (1997) The ammonia monooxygenase structural gene *amoA* as a functional marker: molecular fine-scale analysis of natural ammonia-oxidizing populations. Appl Environ Microbiol 63:4704–4712. https://doi.org/10.1128/aem.63.12.4704-4712.1997
20. Throbäck IN, Enwall K, Jarvis Å, Hallin S (2004) Reassessing PCR primers targeting *nirS*, *nirK* and *nosZ* genes for community surveys of denitrifying bacteria with DGGE. FEMS Microbiol Ecol 49:401–417. https://doi.org/10.1016/j.femsec.2004.04.011

21. Schloss PD, Westcott SL, Ryabin T, Hall JR, Hartmann M, Hollister EB, Lesniewski RA, Oakley BB, Parks DH, Robinson CJ, Sahl JW, Stres B, Thallinger GG, Van Horn DJ, Weber CF (2009) Introducing mothur: open-source, platform-independent, community-supported software for describing and comparing microbial communities. Appl Environ Microbiol 75:7537–7541. https://doi.org/10.1128/AEM.01541-09
22. Tripathi BM, Kim HM, Jung JY, Nam S, Ju HT, Kim M, Lee YK (2019) Distinct taxonomic and functional profiles of the microbiome associated with different soil horizons of a moist tussock tundra in Alaska. Front Microbiol 10:1442. https://doi.org/10.3389/fmicb.2019.01442
23. Huse SM, Welch DM, Morrison HG, Sogin ML (2010) Ironing out the wrinkles in the rare biosphere through improved OTU clustering. Environ Microbiol 12:1889–1898. https://doi.org/10.1111/j.1462-2920.2010.02193.x
24. Rognes T, Flouri T, Nichols B, Quince C, Mahé F (2016) VSEARCH: a versatile open source tool for metagenomics. PeerJ 4:e2584. https://doi.org/10.7717/peerj.2584
25. Kim O-S, Cho Y-J, Lee K, Yoon S-H, Kim M, Na H, Park S-C, Jeon YS, Lee J-H, Yi H, Won S, Chun J (2012) Introducing EzTaxon-e: a prokaryotic 16S rRNA gene sequence database with phylotypes that represent uncultured species. Int J Syst Evol Microbiol 62:716–721. https://doi.org/10.1099/ijs.0.038075-0
26. Westcott SL, Schloss PD (2017) OptiClust, an improved method for assigning amplicon-based sequence data to operational taxonomic units. mSphere 2:e00073–17. https://doi.org/10.1128/mSphereDirect.00073-17
27. Oksanen J, Simpson GL, Blanchet FG, Kindt R, Legendre P, Minchin PR, O'Hara RB, Solymos P, Stevens MH, Szoecs E, Wagner H, Barbour M, Bedward M, Bolker B, Borcard D, Carvalho G, Chirico M, De Caceres M, Durand S, Evangelista HBA, FitzJohn R, Friendly M, Furneaux B, Hannigan G, Hill MO, Lahti L, McGlinn D, Oullette M-H, Ribero Cunha E, Smith T, Stier A, ter Braak CJF, Weedon J (2022) Vegan: community ecology package, R package ver. 2.6-4. https://CRAN.R-project.org/package=vegan
28. Anderson M, Gorley RN, Clarke RK (2008) Permanova+ for PRIMER: guide to software and statistical methods. PRIMER-E Ltd., Plymouth, UK
29. Clarke KR, Gorley RN (2006) PRIMER, 6th edn. PRIMER-E Ltd., Plymouth, UK
30. Zhang K, Cheng X, Shu X, Liu Y, Zhang Q (2018) Linking soil bacterial and fungal communities to vegetation succession following agricultural abandonment. Plant Soil 431:19–36. https://doi.org/10.1007/s11104-018-3743-1
31. Knelman JE, Schmidt SK, Graham EB (2021) Cyanobacteria in early soil development of deglaciated forefields: dominance of non-heterocystous filamentous cyanobacteria and phosphorus limitation of N-fixing Nostocales. Soil Biol Biochem 2021:108127. https://doi.org/10.1016/j.soilbio.2020.108127
32. Cano-Díaz C, Maestre FT, Wang J, Li J, Singh BK, Ochoa V, Gozalo B, Delgado-Baquerizo M (2022) Effects of vegetation on soil cyanobacterial communities through time and space. New Phytol 234:435–448. https://doi.org/10.1111/nph.17996
33. Manpoong C, De Mandal S, Bangaruswamy DK, Perumal RC, Benny J, Beena PS, Ghosh A, Kumar NS, Tripathi SK (2020) Linking rhizosphere soil biochemical and microbial community characteristics across different land use systems in mountainous region in Northeast India. MetaGene 23:100625. https://doi.org/10.1016/j.mgene.2019.100625
34. Wood SA, Gilbert JA, Leff JW, Fierer N, D'Angelo H, Bateman C, Gedallovich SM, Gillikin CM, Gradoville MR, Mansor P, Massmann A, Yang N, Turner BL, Brearley FQ, McGuire KL (2017) Consequences of tropical forest conversion to oil palm on soil bacterial community and network structure. Soil Biol Biochem 112:258–268. https://doi.org/10.1016/j.soilbio.2017.05.019
35. Falkowski PG, Fenchel T, Delong EF (2008) The microbial engines that drive Earth's biogeochemical cycles. Science 320:1034–1039. https://doi.org/10.1126/science.1153213
36. Fierer N, Strickland MS, Liptzin D, Bradford MA, Cleveland CC (2009) Global patterns in belowground communities. Ecol Lett 12:1238–1249. https://doi.org/10.1111/j.1461-0248.2009.01360.x

37. Kallenbach C, Grandy AS (2011) Controls over soil microbial biomass responses to carbon amendments in agricultural systems: a meta-analysis. Agric Ecosyst Environ 144:241–252. https://doi.org/10.1016/j.agee.2011.08.020
38. Sailo AS, Lalmuankimi C (2012) Conservation of Reiek Tlang: a brief assessment of the adverse effect of vehicles and a plea for its cessation. Sci Vis 12:2–7
39. Brearley FQ, Thomas AD (2015) Land-use change impacts on soil processes: tropical and savannah environments. CABI, Wallingford, UK
40. Nath PC, Nath AJ, Reang D, Lal R, Das AK (2021) Tree diversity, soil organic carbon lability and ecosystem carbon storage under a fallow age chronosequence in North East India. Environ Sustain Indic 10:100122. https://doi.org/10.1016/j.indic.2021.100122
41. Sahoo UK, Singh SL, Gogoi A, Kenye A, Sahoo SS (2019) Active and passive soil organic carbon pools as affected by different land use types in Mizoram, Northeast India. PLoS ONE 14:e0219969. https://doi.org/10.1371/journal.pone.0219969
42. Li W, Zhang Y, Mao W, Wang C, Yin S (2020) Functional potential differences between *Firmicutes* and *Proteobacteria* in response to manure amendment in a reclaimed soil. Can J Microbiol 66:689–697. https://doi.org/10.1139/cjm-2020-0143
43. Rocca JD, Lennon JT, Evans SE (2014) Relationships between protein-encoding gene abundance and corresponding process are commonly assumed yet rarely observed. ISME J 9:1693–1699. https://doi.org/10.1038/ismej.2014.252
44. Sharma PK, Sharma V, Sharma S, Bhatia G, Singh K, Sharma R (2019) Comparative metatranscriptome analysis revealed broad response of microbial communities in two soil types, agriculture versus organic soil. J Genet Eng Biotechnol 17:6. https://doi.org/10.1186/s43141-019-0006-3
45. Han L-L, Wang Q, Shen J-P, Di HJ, Wang J-T, Wei W-X, Fang Y-T, Zhang L-M, He J-Z (2019) Multiple factors drive the abundance and diversity of the diazotrophic community in typical farmland soils of China. FEMS Microbiol Ecol 95:fiz113. https://doi.org/10.1093/femsec/fiz113
46. Aboim MCR, Coutinho HLC, Peixoto RS, Barbosa JC, Rosado AS (2008) Soil bacterial community structure and soil quality in a slash-and-burn cultivation system in Southeastern Brazil. Appl Soil Ecol 38:100–108. https://doi.org/10.1016/j.apsoil.2007.09.004

Screening of Rhizosphere Actinomycetes from Shifting Cultivation Crops for Pigment Production

Marcy D. Momin and Shri Kant Tripathi

Abstract Among the microorganisms, actinomycetes can produce various natural pigments that have antimicrobial compounds and antioxidant properties. In this regard, pigment-producing actinomycete strains could be an essential source for producing antimicrobial agents to develop bioinoculants in agricultural programs. This study aimed to isolate actinomycetes from the rhizosphere of crop plants under shifting cultivation in Mizoram, North-east India, capable of producing pigments. Rhizosphere soil samples were collected and actinomycetes were isolated from the soil by serial dilution. Purified isolates were identified by their plate colony nature and microscopical morphology, and characterised by Bergey's Manual of Determinative Bacteriology. A total of ten actinomycete taxa were recorded that produced various pigments. This study suggests that rhizospheric actinomycetes from major crop plants under shifting cultivation in Mizoram might produce bioactive compounds responsible for pigment production.

Keywords Actinomycetes · Crops · Pigment production · Rhizosphere · Shifting cultivation

1 Introduction

Pigments of various colours are produced by microorganisms to shield the cells from the damaging effects of light rays of the visible and near-ultraviolet range. Previous researchers have found that the pigments like carotenoids, flavins, melanin, prodigiosin and violacein showed distinct antimicrobial effects against many pathogenic bacteria [1–6]. Furthermore, pigments are of interest to many industries as natural pigments are utilised in foodstuffs, dyes, cosmetics and pharmaceuticals [7, 8]. These natural sources of colourants or dyes derived from microbes are purported to be non-toxic, non-carcinogenic and biodegradable [9]. Furthermore, some actinomycetes

M. D. Momin (✉) · S. K. Tripathi
Department of Forestry, Mizoram University, Aizawl, Mizoram 796004, India
e-mail: mominmarcy@gmail.com

S. K. Tripathi and F. Q. Brearley (eds.), *Shifting Cultivation Systems*,
https://doi.org/10.1007/978-3-031-70388-1_13

are capable of producing melanin that protects the living organisms from ultraviolet radiation, and they have antioxidant properties [10, 11].

Actinomycetes are gram-positive bacteria, ubiquitous in nature with a high G+C content in their DNA. While actinomycetes are bacteria, they resemble fungi in their morphology in that they form branching filaments or hyphae and asexual spores. Actinomycetes are known as a potential source of many bioactive compounds, and they also have tremendous capability to produce natural pigments [12]. It has been estimated that approximately one-third of the thousands of naturally occurring antibiotics have been obtained from the actinomycete genus *Streptomyces* [13]. Hence, the isolation of pigment-producing actinomycetes has been a prime area of current research. Our main objective in the present study was to isolate pigment-producing rhizosphere actinomycetes from the crop plants under shifting cultivation in the state of Mizoram, North-east India and screen them for pigment production.

2 Materials and Methods

Soil samples were collected from the rhizosphere of six common crop plants *Capsicum annuum* (Solanaceae), *Dioscorea oppositifolia* (Dioscoreaceae), *Oryza sativa* (Poaceae), *Solanum melongena* (Solanaceae) *Vigna mungo* (Fabaceae) and *Zea mays* (Poaceae) grown under shifting cultivation (jhum) around the village of Reiek, Mamit district, Mizoram (23° 41′ N, 92° 36′ E at an elevation of 1465 m) in 2018. Soil was collected from three different shifting cultivation fields with three replicate samples from each of the six crop species. Collected soil was dried at room temperature overnight and then serially diluted in sterile deionised water for isolation of actinomycetes [14]. A 100 μl diluted aliquot was inoculated on three selected media by the spreading method with a sterile L-spreader and the inoculated plates were then incubated at 28 °C for 1–4 weeks. Selected media were starch casein agar (SCA), cross streak media (CSM), and international Streptomyces agar (ISP2). Dry, sticky, fuzzy, filamentous, and both aerial and substrate mycelium were recorded morphologically for the actinomycete colonies. Several times, selected colonies were sub-cultured by streaking methods until a pure culture was obtained. These isolated pure single cultures were then studied for their morphological characteristics according to Bergey's Manual of Determinative Bacteriology [15]. All the identified isolates were preserved in 25% glycercol stock at – 80 °C for further use.

3 Results and Discussion

In the present study, a total of ten rhizospheric actinomycete taxa were found to produce pigmented colonies during their normal growth. Mostly, they showed slow growth. Their morphology was very sticky-hard, irregular, convex in size and shape, filamentous, and powdery in the aerial mycelium (Fig. 1). They produced pigments that were brown, light-yellow, bright-yellow or orange-yellow (Fig. 1). Out of ten identified isolates, eight showed a similar morphology to *Streptomyces* sp., one was similar to *Micromonospora* sp., and a final isolate resembled *Nocardia* sp. (Table 1). The resistance of actinomycetes to stress conditions may lead to the production of various bioactive compounds including pigments. *Streptomyces* sp. is common actinomycete genus and was the most prevalent genus in the present study. *Streptomyces* is an excellent producer of bioactive metabolites and serves as an important source for the discovery of novel bioactive products [16]. Besides *Streptomyces* sp., other genera like *Micromonospora* sp. and *Nocardia* sp. are rare actinomycetes [17] that have also been revealed to be a unique source of natural products [18].

Previous researchers have reported that actinomycetes produce various pigments on the growth media and showed the production of antibiotics by the rarer actinomycetes *Micromonospora* and *Nocardia* [19, 20]. This group of organisms was considered an important soil component that can degrade recalcitrant compounds and produce dark black to brown pigments that contribute to the dark colour of soil humus [21]. An earlier study was conducted to determine actinomycete antibacterial and antifungal potential isolated from soil samples with actinomycete isolates showing various pigments such as pink, yellow, orange, red, and brown, with 30% of the isolates demonstrating antimicrobial activities [22]. In another earlier study, pigment-producing actinomycetes were isolated and tested for antimicrobial activity,

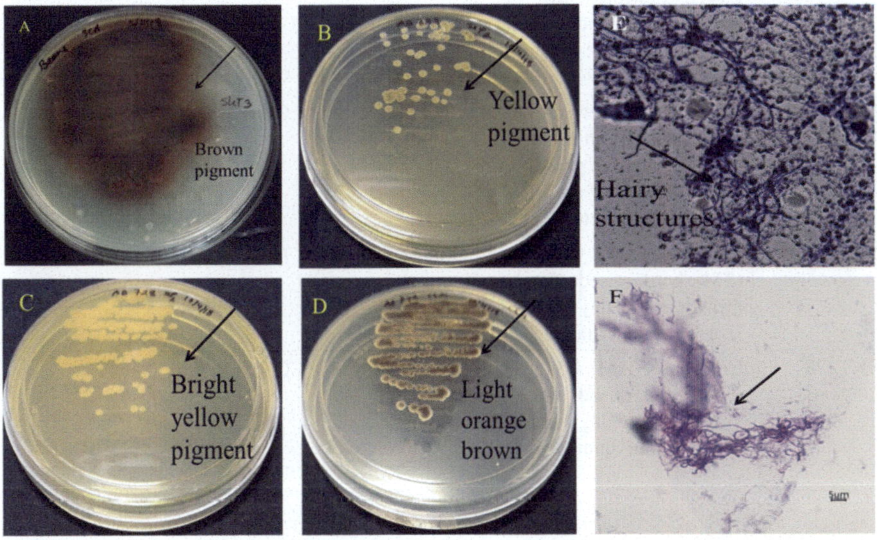

Fig. 1 Representatives of actinomycete isolates: **a** SKT3 showed the capability of producing brown pigment. **b** SKT729 showing the capability to produce light-yellow pigment. **c** SKT729 showing the capability to produce bright yellow pigment. **d** SKT742 showing the capability to producing light-brown pigment. **e, f** Microscopic images of actinomycetes. All of the actinomycete isolates are from shifting cultivation fields of Mizoram, North-east India

Table 1 Morphological characteristics of the actinomycete isolates from shifting cultivation fields of Mizoram, North-east India

Isolate code	Colony morphology	Pigment colour	Media	Taxa
SKT3	Brown powdery (AM), deep brown (SM); sticky, raised; growth in 5–7 days; colony size 0.8 mm in diameter	Brown	SCA	*Streptomyces* sp.
SKT728	Colourless (AM), pure yellow (SM); sticky, hard, raised; growth in 5–7 days; colony size 1 mm in diameter	Bright yellow	ISP2	*Streptomyces* sp.
SKT729	Colourless (AM), green circle (SM); sticky, hard, convex; growth in 5–7 days; colony size 1 mm in diameter	Light yellow	ISP2	*Streptomyces* sp.
SKT732	Colourless (AM), green circle (SM); sticky, hard, convex; growth in 5–7 days; colony size 1 mm in diameter	Deep yellow	CSM	*Streptomyces* sp.
SKT742	Off-white with grey spots (AM), off-white (SM); sticky, convex and irregular; growth in 5 days; colony size 1.5 mm in diameter	Yellow	CSM	*Streptomyces* sp.
SKT759	Off-white powdery (AM), yellow (SM); soft-sticky, flat, filamentous; growth in 3–5 days; colony size 1.5 mm in diameter	Turmeric yellow	CSM	*Micromonospora* sp.
SKT774	Off-white (AM), yellow (SM); hard, raised, circular; growth in 5–7 days; colony size 1.2 mm in diameter	Orange-yellow	ISP2	*Streptomyces* sp.
SKT778	Colourless (AM), colourless (SM); hard, raised, entire; growth in 5–7 days; colony size 1 mm in diameter	Light yellow	CSM	*Streptomyces* sp.

(continued)

Table 1 (continued)

Isolate code	Colony morphology	Pigment colour	Media	Taxa
SKT782	Off-white (AM), yellow-orange (SM); soft, fuzzy, raised, circular; growth in 3–5 days; colony size 0.5 mm in diameter	Light orange	SCA	*Nocardia* sp.
SKT788	Off-white with dark-grey spot (AM), yellow with green spot (SM); sticky-hard, raised, entire; growth in 5–7 days; colony size 1 mm in diameter	Dark yellow	ISP2	*Streptomyces* sp.

AM = Ariel mycelium, SM = substrate mycelium

which showed that all extracted pigments inhibited the growth of *Bacillus subtilis, Escherichia coli, Salmonella typhi* and *Staphylococcus aureus* [23, 24].

4 Conclusions

Shifting cultivation is the primary agricultural system in North-east India. The major crops under shifting cultivation have a large rhizosphere microbial population, including actinomycetes which have a natural pigment-producing ability. These filamentous bacteria are reported as an ecologically diverse group. The results of our study indicate the natural pigment-producing actinomycete may also produce antimicrobial compounds. The antimicrobial potential of extracellular metabolites produced by the pigment-producing actinomycetes may be further studied for its future use as a biocontrol agent or bioinoculant in sustainable shifting agriculture management.

References

1. Cang S, Sanada M, Johdo O, Ohta S, Nagamatsu Y, Yoshimoto A (2000) High production of prodigiosin by *Serratia marcescens* grown on ethanol. Biotechnol Lett 22:1761–1765. https://doi.org/10.1023/A:1005646102723
2. Kim D, Lee JS, Park YK, Kim JF, Jeong H, Oh T-K, Kim BS, Lee CH (2007) Biosynthesis of antibiotic prodiginines in the marine bacterium *Hahella chejuensis* KCTC 2396. J Appl Microbiol 102:937–944. https://doi.org/10.1111/j.1365-2672.2006.03172.x
3. Mortensen A (2006) Carotenoids and other pigments as natural colorants. Pure Appl Chem 78:1477–1491. https://doi.org/10.1351/pac200678081477
4. Pantanella F, Berlutti F, Passariello C, Sarli S, Morea C, Schippa S (2007) Violacein and biofilm production in *Janthinobacterium lividum*. J Appl Microbiol 102:992–999. https://doi.org/10.1111/j.1365-2672.2006.03155.x

5. Vasanthabharathi V, Lakshminarayanan R, Jayalakshmi S (2013) Melanin production from marine *Streptomyces*. Afr J Biotechnol 10:11224–11234. https://doi.org/10.5897/AJB11.296
6. Venil CK, Lakshmanaperumalsamy P (2009) An insightful overview on microbial pigment, prodigiosin. Electron J Biol 5:49–61
7. Malik K, Tokkas J, Goyal S (2012) Microbial pigments: a review. Int J Microb Resour Technol 1:361–365
8. Venil CK, Zakaria ZA, Ahmad WA (2013) Bacterial pigments and their applications. Process Biochem 48:1065–1079. https://doi.org/10.1016/j.procbio.2013.06.006
9. Sinha S, Choubey S, Kumar AA, Bhosale P (2017) Identification, characterization of pigment producing bacteria from soil and water and testing of antimicrobial activity of bacterial pigments. Int J Pharm Sci Rev Res 42:119–124
10. Dastager SG, Li W-J, Dayanand A, Tang S-K, Tian X-P, Zhi X-Y, Xu L-H, Jiang C-L (2006) Separation, identification and analysis of pigment (melanin) production in Streptomyces. Afr J Biotechnol 5:1131–1134
11. Frases S, Salazar A, Dadachova E, Casadevall A (2007) *Cryptococcus neoformans* can utilize the bacterial melanin precursor homogentisic acid for fungal melanogenesis. Appl Environ Microbiol 73:615–621. https://doi.org/10.1128/AEM.01947-06
12. Selim MSM, Abdelhamid SA, Mohamed SS (2021) Secondary metabolites and biodiversity of actinomycetes. J Genet Eng Biotechnol 19:72. https://doi.org/10.1186/s43141-021-00156-9
13. Bérdy J (2005) Bioactive microbial metabolites: a personal view. J Antibiot 58:1–26. https://doi.org/10.1038/ja.2005.1
14. Sengupta S, Pramanik A, Ghosh A, Bhattacharyya M (2015) Antimicrobial activities of actinomycetes isolated from unexplored regions of Sundarbans mangrove ecosystem. BMC Microbiol 15:170. https://doi.org/10.1186/s12866-015-0495-4
15. Holt JG, Krieg NR, Sneath PHA, Staley JT, Williams ST (2000) Bergey's manual of determinative bacteriology, 9th edn. Lippincott Williams and Wilkins, Philadelphia, USA
16. Ryan RP, Germaine K, Franks A, Ryan DJ, Dowling DN (2008) Bacterial endophytes: recent developments and applications. FEMS Microbiol Lett 278:1–9. https://doi.org/10.1111/j.1574-6968.2007.00918.x
17. Qin S, Chen H-H, Zhao G-Z, Li J, Zhu W-Y, Xu L-H, Jiang J-H, Li W-J (2012) Abundant and diverse endophytic actinobacteria associated with medicinal plant *Maytenus austroyunnanensis* in Xishuangbanna tropical rainforest revealed by culture-dependent and culture-independent methods. Environ Microbiol Rep 4:522–531. https://doi.org/10.1111/j.1758-2229.2012.00357.x
18. Subramani R, Sipkema D (2019) Marine rare actinomycetes: a promising source of structurally diverse and unique novel natural products. Mar Drugs 17:249. https://doi.org/10.3390/md17050249
19. Manandhar S, Sharma S (2017) Practical approach to microbiology, 3rd edn. National Book Centre, Kathmandu, Nepal
20. Narain S, Abhay S, Sen S, Mondal S, Manigandan N, Yogesh BJ (2014) Study and classification of red pigmented actinomycetes from the Bannerghatta National Park soil. J Int Acad Res Multidiscip 2:347–358
21. Rathore P, Ara Y, Joshi A (2012) An insight into agricultural properties of actinomycetes. Int J Res BioSci 1:7–12
22. Rana S, Salam MD (2014) Antimicrobial potential of actinomycetes isolated from soil samples of Punjab, India. J Microbiol Exp 1:63–68. https://doi.org/10.15406/jmen.2014.01.00010
23. Tandale A, Khandagale M, Palaskar R, Kulkarni S (2018) Isolation of pigment producing actinomycetes from soil and screening their antibacterial activities against different microbial isolates. Int J Curr Res Life Sci 7:2397–2402
24. Sapkota A, Thapa A, Budhathoki A, Sainju M, Shreshtha P, Aryal S (2020) Isolation, characterization, and screening of antimicrobial-producing actinomycetes from soil samples. Int J Microbiol 2716584. https://doi.org/10.1155/2020/2716584

Sustaining Shifting Agricultural Practices in North-East India Through Beneficial Soil Microbes

Gajanan T. Mehetre, Purbajyoti Deka, Bhim Pratap Singh, and Kalidas Upadhyaya

Abstract The shifting agricultural system represents the prime approach for various communities in North-east India for food production. In shifting cultivation, an area of forested land is slashed and burnt, making it ready for crop production for two to three years and then it is left as fallow land. This process typically involves the natural recovery of the soil after cropping. Therefore, soil fertility depends to a large extent on the biochemical processes performed by beneficial soil microorganisms and their functional diversity. Currently shifting cultivation is practiced with a reduced fallow period that has a severe impact on soil microbial diversity and soil erosion that ultimately causes a reduction in crop productivity. A suitable solution to such agricultural practices is important for environmental sustainability and to increase the economic well-being of the indigenous farmers. Currently, the low soil fertility problem is being tackled by various amendments like the use of organic and inorganic fertilisers in soils. The use of organic fertilisers that contain beneficial microbes is more sustainable and holds promise for agricultural productivity. One important aspect is to better understand the composition of microbial communities from soils under shifting cultivation along with their beneficial properties. A few basic studies have been reported that have explored the different microbial taxa from soils under shifting cultivation fields in North-east India. However, studies aiming towards elucidating microbial interactions and their stimulatory response on the host plant remain largely unreported from this region. Successful manipulation of long-term and persistent plant beneficial microbial communities in shifting cultivation will benefit the farmers of this region. However, although the current status of knowledge demonstrates the potential to improve crop productivity using beneficial microbes,

G. T. Mehetre · P. Deka
Department of Biotechnology, Mizoram University, Aizawl 796004, Mizoram, India

B. P. Singh
Department of Agriculture and Environmental Sciences, National Institute of Food Technology Entrepreneurship and Management (NIFTEM), Industrial Estate, Kundli, Sonepat, Haryana 131028, India
e-mail: bpsingh@niftem.ac.in

K. Upadhyaya (✉)
Department of Forestry, Mizoram University, Aizawl, Mizoram 796004, India
e-mail: kumzu70@gmail.com

S. K. Tripathi and F. Q. Brearley (eds.), *Shifting Cultivation Systems*,
https://doi.org/10.1007/978-3-031-70388-1_14

this area needs more research attention. In this context, this chapter discusses the studies related to microbial diversity and beneficial microbes under shifting cultivation and their impact on soil health and agricultural productivity that can sustain shifting agricultural practices in North-east India.

Keywords Microbial diversity · North-east India · Shifting cultivation · Soil fertility · Sustainability

1 Introduction

Shifting cultivation, also known as slash-and-burn agriculture, is an age-old agricultural system practiced for crop production in the hilly parts of the North-eastern region of India [1]. Shifting cultivation involves a cyclic process of cropping and subsequent fallowing of land. In this system, steeply sloping land is cleared by cutting natural vegetation that is burned in order to prepare a piece of land for cultivation. After two or three years of cultivation, the land is left fallow for several years for restoration of the vegetation biomass [2]. The length of the fallow period is typically for more than a decade (15–20 years) to establish natural vegetation cover that can improve soil fertility. However, in recent years, the fallow period has shortened to less than five years. As a direct consequence of the shortening of the fallow period, it typically results in the accentuation of soil erosion and nutrient losses from the topsoil [3]. Because of such negative impacts on environmental sustainability, the shifting cultivation agricultural system faces several criticisms [4, 5]. Although critical perspectives on shifting cultivation have been elaborated by several policies, shifting agricultural practices nevertheless continue in North-east India because of its cultural importance to the indigenous population. Naturally, the shifting cultivation system could be promoted sustainably by recycling farm wastes (e.g. manure and crop residues) and diversification of crop plants along with the use of plant growth-promoting bacteria [6–8].

In this specific context, soil microbes perform a crucial role in improving soil health and promoting plant growth. Soil microbial diversity and composition depends on various factors like temperature, moisture, pH, and soil type. In addition, plant root exudates released into the soil also act as a chemical signal for associated microbial taxa that form the rhizome-microbiome of the plant [9]. The plant rhizosphere is the complex interface between plant roots and soil where interactions among diverse microbial taxa exert a profound influence on soil biochemical properties and are directly or indirectly involved in plant growth promotion and development by stimulating nutrient acquisition and helping plants to tolerate abiotic stresses [10]. Many of the current insights into interactions and processes in the rhizosphere have emerged from studies on agricultural crop plants [11]. In agricultural systems, rhizospheric microbial taxa carry out the decomposition of plant residues. Microbial activities and other physicochemical agents have an enormous impact on the chemical composition of organic matter. In most cases, the decomposition of organic matter is performed by heterotrophic microorganisms [12]. Therefore, studies on native microbial communities from shifting cultivation areas of North-east India are important. As a consequence, in recent years, studies have focused on soil microbial diversity of such cultivated lands to keep at par with the enormous amount of effort that is being made worldwide by microbial ecologists to identify microorganisms in numerous environmental samples [13]. However, progress has lacked momentum in understanding their composition, diversity, and ecology from the regions where shifting cultivation is prevalent. In recent years, a few studies have documented total numbers of microorganisms, biomass, respiration rates, and enzyme activities, but with little attention being paid towards whole microbial communities and their function [10, 14, 15]. To sustain shifting cultivation, promoting the use of bio-fertilisers and other microbes beneficial to plant growth could serve as one of the viable options. In this chapter, studies based on microbial diversity and their applicability for plant growth promotion are discussed. In addition, relevant studies related to the impact of soil amendments with organic and inorganic fertilisers on soil microbial diversity are also discussed and evaluated.

2 Shifting Cultivation and the Rhizosphere

The rhizosphere represents a complex association of microbial communities with plant roots and soil. The microbial composition of the rhizosphere represents a biological niche for diverse micro-organisms because of organic matter derived from plant parts and root exudates. The complex association of plants, microbes, and soil in the rhizosphere is chiefly driven by plant composition, soil types, and the associated environmental conditions [11]. The rhizosphere has crucial importance in plant health and the development of vegetative cover during ecological restoration. Microbial taxa represent the first colonisers in the slashed-and-burned land after the fallow period in shifting cultivation. Therefore, rhizosphere microbial composition and their activities are important for vegetation development and successful recovery during vegetation reestablishment in the fallow period [16]. As more species colonise the rhizosphere, the biochemical activities occurring in that soil enriches its nutrient content. Microbial species efficiently utilise plant residues for their growth; however, the rate of decomposition is dependent on the physicochemical environment and nature of the microbial community in the soil and the amount and nature of the biochemical composition of organic material [17].

Heterotrophic microorganisms like fungi and bacteria carry out the biochemical decomposition of organic material [18]. Fungal species associated with plant roots are involved in the decomposition of the more recalcitrant organic material like lignin and cellulose. Bacteria also constitute the major microbial fraction in the rhizosphere soil. Bacteria are typically the first organisms to sense and respond to chemical and physical changes in their environment [19]. Most of the bacterial species are known to produce a variety of extracellular hydrolytic enzymes responsible for the decomposition of organic matter. However, the process of decomposition is performed by microbial community succession on a natural substratum that sequentially releases different organic and inorganic nutrients. Succession of the microbial communities also brings about changes in the chemical composition of the organic matter that, in turn, regulates the microbial composition of late colonising microorganisms [20]. Therefore, it is important to understand the magnitude of the rhizosphere effect on microbial composition and their properties. Such studies have mostly been reported on natural ecosystems and conventional agricultural soils [21]. Given the substantial contribution of the rhizosphere microbial communities to plant growth and development, exploration and utilisation of such microorganisms could be a key to unfolding agricultural constraints on shifting cultivation and achieving increased productivity. However, so far, little is known about the rhizosphere soil microbial communities under shifting agricultural systems. A few studies have reported isolation and characterisation of bacteria, fungi and other microorganisms under shifting cultivation from different states of North-east India (Table 1). Most of the reported studies showed that the isolated microorganisms possess plant growth-promoting properties and were beneficial for plant health which is discussed further here.

3 Beneficial Microbes and Plant Growth Promoting Rhizobacteria (PGPR) in Shifting Cultivation

In the agricultural system, microbes play important roles in improving soil health and help in plant growth promotion. In addition, microbes also help in bio-control of plant pathogenic diseases [46]. In the rhizosphere, the composition of the microbial community is different to that of the surrounding soil. Microbial species have plant growth-promoting properties such as auxin synthesis, siderophore production, and phosphate-solubilisation, etc. [8]. Various studies conducted on rhizosphere soils and endophytes of plants have reported the presence of beneficial bacterial communities that have various plant growth-promoting traits and could have biotechnological applications as, for example, bio-control agents [47, 48]. However, such beneficial microbial communities have rarely been investigated under shifting cultivation, especially from North-east India. A few such studies reported the isolation and screening for PGPR traits or other activities from North-east India are shown in Table 1.

Various studies have explored the diversity of bacteria and fungi from fields under shifting cultivation in North-east India. Most of these studies reported the isolation of diverse microbial taxa with various properties like PGPR and bio-control activities. Momin et al. [37] studied the rhizosphere soils of major traditional crop plants under shifting cultivation in Mizoram for their salt tolerance and PGPR activities. The rhizosphere soils collected from two locations were used for isolating actinomycetes using culture media. They reported the isolation of around 26 actinomycete strains, out of which, five strains had salt tolerance and one strain had multiple PGPR traits. In total, eight strains were positive for ammonia production. The strain with PGPR properties was identified as *Micromonospora auratinigra* based on 16SrRNA

Table 1 Microbial diversity, beneficial microbes and plant growth promoting rhizobacteria studied in shifting cultivation agro-ecosystems of North-east India

Study location(s)	Aim of the study	Dominant microbial taxa/strains observed	Application in agriculture	Reference
Chandel, Churachandpur, Tamenglong and Ukhrul districts of Manipur, India	Assessment of the impact of biofertilisers and micro-dosing of NPK on productivity of rice, maize, groundnut and soybeans in shifting cultivation lands	Not performed	Application of *Azotobacter* and phosphate-solubilising bacteria and micro-dosing of NPK for productivity enhancement in shifting cultivation fields	Baishya et al. [22]
Shifting cultivation fields, North-east India	Isolation and screening for PGPR bacterial strains	*Bacillus cereus*, *B. methylotrophicus*, *Curtobacterium oceanosedimentum*, *Metarhizium pinghaense*, *Penicillium stratisporum* and *P. virgatum*	Production of indole compounds, siderophores, ammonia, phosphate-solubilisation, and catalase activity	Banerjee et al. [9]
Mokokchung district, Nagaland, India	Isolation of endophytic bacterial species from Broom grass (*Thysanolaena maxima*; Poaceae) rhizospheric soil	Various bacterial taxa	Production of pectinase and cellulase, IAA-like substances, ACC-deaminase, N_2-fixation, solubilisation of inorganic P	Deka et al. [23]
Imphal valley, Manipur, India	Characterisation of rhizosphere bacterial diversity	*Burkholderia* sp. and *Pseudomonas* sp.	Antagonistic activity against fungal pathogens	Devi et al. [24]
Khonoma, Nagaland, India	Determination of soil bacterial taxa	*Kosakonia sacchari*	Diazotrophic potential, positive for *nifH* genes and ammonia production	Giri [25]
Longkhum and Ungma villages in Mokokchung, Nagaland, India	Evaluation of the performance of native PGPRs on rice yield enhancement	*Azospirillum oryzae*, *Bacillus cereus* and *Pseudomonas fluorescens*	Native PGPRs isolated from the rice rhizosphere for productivity enhancement in shifting cultivation fields	Giri et al. [26]

(continued)

Table 1 (continued)

Study location(s)	Aim of the study	Dominant microbial taxa/strains observed	Application in agriculture	Reference
Muallungthu and Tachhip, Mizoram, India	Rhizosphere effects of bamboo (*Melocanna baccifera*; Poaceae)	Not performed	Fallow management in the shifting cultivation areas	Hauchhum and Tripathi [27]
Serchhip district, Mizoram, India	Isolation, identification and characterisation of phosphate-solubilising bacteria from shifting cultivation fields	*Alcaligenes faecalis*, *A.* sp., *Bacillus cereus*, *Ba. paramycoides*, *Ba. subtilis*, *Ba. thuringiensis*, *Ba.* sp., *Burkholderia cepacia*, *Bu. gladioli*, *Bu. glumae*, *Bu. vietnamese*, *Bu.* sp., *Pseudomonas bifermentans* and *P. oryzihabitans*	Application of native phosphate-solubilising bacteria with PGPR characteristics for increasing crop productivity in shifting cultivation areas	Hnamte et al. [28]
Lobi, Itanagar, Arunachal Pradesh, India	Identification and characterisation of fungal strains from burned plots	*Aspergillus* sp., *Cladosporium* sp., *Paecilomyces* sp., *Penicillium* sp. and *Trichoderma* sp.	Ligninolytic activity, phosphate-solubilisation and several properties with respect to bio-control including production of lytic enzymes, volatile and diffusible antimicrobials, and ammonia	Jain et al. [29]
Lahdoigarh, Jorhat, Assam, India	Isolation and identification PGPR strains	*Bacillus cereus*, *Chromobacterium violaceum*, *Pseudomonas rhodesiae* and *Streptomyces luteireticuli*	Bio-formulation as effective growth and nutrient content enhancer for agricultural crops	Kalita et al. [30]
Manipur, India	Isolation of keratinase producing actinobacterial species from rhizosphere of tomato (*Solanum lycopersicum*; Solanaceae)	*Streptomyces cinereoruber*, *S. ederensis*, *S. polychromogenes*, *S. pratensis*, *S. tanashiensis*, *S. tunisiensis*, *S. vinaceus* and *S. virginiae*	Strains with phyto-stimulating and bio-control properties	Kshetri et al. [31]

(continued)

Table 1 (continued)

Study location(s)	Aim of the study	Dominant microbial taxa/strains observed	Application in agriculture	Reference
Sacred groves of Meghalaya, India	Bioprospecting of plant growth promoting *Bacillus* species	*Bacillus* sp., *Lysinibacillus* sp., *Paenibacillus* sp. and *Viridibacillus* sp.	Positive strains with multiple PGPR genetic traits such as biosynthesis of acid phosphatase, indolepyruvate decarboxylase, ACC-deaminase and siderophore biosynthesis protein	Lyngwi et al. [32]
Changlang, Arunachal Pradesh, India	Decomposition of dominant weeds in a shifting cultivation system	*Chrysosporium merdarium*, *Cladosporium cladosporioides*, *C. macrocarpum*, *Geotrichum candidum*, *Metarhizium* sp., *Monilia* sp., *Mortierella elongata*, *Penicillium brevicompactum*, *P. canescens*, *Nigrospora sphaerica* and *Verticillium* sp.	Not performed	Majumder et al. [33]
Oil palm plantation and forest, Mizoram, India	Analysing soil biochemical and microbial characteristics	*Acidobacteria* sp., *Pseudomonadota* sp. and *Verrucomicrobiota* sp.	Not performed	Manpoong et al. [10]
Reiek and Tanhril, Mizoram, India	Isolation of rhizospheric actinomycetes from major crop plants	*Micromonospora* sp. and *Streptomyces* sp.	PGPR strains	Momin [34]
Mizoram University campus, Aizawl, Mizoram, India	Isolation of endophytic bacteria and studies of their properties from *Ageratum conyzoides* and *Mikania micrantha* (both Asteraceae)	*Aneurinibacillus aneurinilyticus*, *Bacillus sonorensis*, *Clostridium frigidicarnis*, *Lysinibacillus fusiformis*, *Micrococcus luteus* and *Saccharopolyspora* sp.	PGPR properties, IAA production, phosphate-solubilisation, ammonia production, cellulose and amylase production	Momin and Ibrahim [35]

(continued)

Table 1 (continued)

Study location(s)	Aim of the study	Dominant microbial taxa/strains observed	Application in agriculture	Reference
Mizoram, India	Isolation of actinomycetes from the rhizosphere soils	Actinomycete species	Strains with phosphate-solubilisation, indole production, siderophore properties	Momin and Tripathi [36]
Reiek and Lengpui, Mizoram, India	Assessment of rhizosphere soils for bacterial diversity	Actinomycete taxa, *Micromonospora aurantinigra*	PGPRs with salt tolerance and ammonia production	Momin et al. [37]
Lobi, Itanagar, Arunachal Pradesh, India	Assess microbial diversity after fire operations in shifting cultivation	*Bacillus clausii*, *B. licheniformis*, *B. megaterium*, *B. subtilis*, *B. thuringiensis*, *Pseudomonas aeruginosa* and *P. stutzeri*	Not performed	Pandey et al. [14]
Mizoram, India	Isolation and screening of cellulose degrading microbes	Not performed	Cellulose degradation	Sangma and Thakuria [38]
Karbi Anglong, Assam, India	Identification of arbuscular mycorrhizas in abandoned land after shifting cultivation	*Glomus aggregatum*, *G. ambisporum*, *G. fulvum* and *G. multicaule*	Not performed	Sarmah [39]
Zaphumi, Lumami and Aotsakili villages in Zunheboto district, Nagaland, India	Evaluation of the response of biofertiliser rhizobium, phosphate-solubilising bacteria and farmyard manure on soybean in shifting cultivation fields	Not performed	Application of biofertiliser rhizobium, phosphate-solubilising bacteria and farmyard manure for increasing soil fertility, and soybean yield in shifting cultivation areas	Sentimenla [40]
Karbi Anglong hills, Assam, India	Exploration of arbuscular mycorrhizal fungal diversity	*Acaulospora* sp., *Ambispora* sp., *Claroideoglomus* sp., *Gigaspora* sp., *Glomus* sp. and *Rhizophagus* sp.	Not performed	Sharmah and Jha [41]

(continued)

Table 1 (continued)

Study location(s)	Aim of the study	Dominant microbial taxa/strains observed	Application in agriculture	Reference
Hmuifang, Mizoram, India	Assessment of soil microbial diversity	Actinomycete species	Not performed	Singh and Tripathi [42]
Papumpare district, Arunachal Pradesh, India	Assessment of species diversity of arbuscular mycorrhizal fungi in shifting cultivation fallows and natural forest soils	Forty-four mycorrhizal fungal species belonging to six genera: *Acaulospora* (6 species), *Enterophospora* (1 species), *Gigaspora* (3 species), *Glomus* (25 species), *Sclerocystis* (2 species) and *Scutellospora* (7 species)	Not performed	Singh et al. [43]
Taphou Naga hills, Manipur, India	Assessment of arbuscular mycorrhizal fungal diversity	*Acaulospora* sp., *Funneliformis* sp., *Glomus* sp., *Rhizophagus* sp., *Sclerocystis* sp. and *Scutellospora* sp.	Not performed	Surendirakumar et al. [44]
Lengpui, Mizoram, India	Evaluation of the impact of fallow length, indigenous soil microbes and rock phosphate on soil fertility and crop productivity	Various microbial inoculants	Microbial inoculum with rock phosphate for soil fertility improvement for rice (*Oryza sativa*; Poaceae) growth in shifting cultivation system	Wapongnungsang et al. [45]

sequencing. This study showed that changes in microbial composition and their properties under shifting cultivation depends on soil physicochemical properties in different crops' rhizospheres. To explore the bacterial diversity associated with Himalayan Alder (*Alnus nepalensis*; Betulacaee) based shifting cultivation in Nagaland, Giri [25] performed the isolation of culturable bacteria identified using 16S rRNA gene sequencing. The key isolated strain was found to be diazotrophic showing the presence of the *nifH* gene and could grow in a N-free medium. This strain was identified as *Kosakonia sacchari* from the shifting cultivation fields.

Isolation of bacterial species with bio-control activity against various plant pathogenic fungi have also been investigated. Devi et al. [24] reported the isolation and characterisation of *Burkholderia cepacia* IBSD-LKBI strain for the control of wilt and damping-off diseases in French beans (*Phaseolus vulgaris*; Fabaceae). This strain exhibited strong activity against fungal pathogens. The treated plants were taller, heavier and had longer roots, and also showed greater disease suppression than in control plants. In addition, the researchers also reported the production of extracellular hydrolytic enzymes like protease, chitinase and β-1,3-glucanase and other PGPR traits like the production of indole acetic acid (IAA) and phosphate-solubilisation. Such microbial taxa and their products can be developed into potential bio-control packages/bio inoculants against soil-borne fungal pathogens.

Symbiotic fungal species in the rhizosphere also play a substantial role in plant health. Arbuscular mycorrhizal fungi, the symbiotic association between Glomeromycotinian fungi and the roots of vascular plants, facilitate the uptake of immobile soil nutrients and perform an important role in the stabilisation of soil aggregates [49]. Jain et al. [29] reported the identification and characterisation of 30 fungal taxa from burned plots under shifting cultivation. The isolated taxa were identified as *Aspergillus* sp., *Cladosporium* sp., *Paecilomyces* sp., *Penicillium* sp., and *Trichoderma* sp. They were found to grow at a broad temperature and pH range and exhibited a range of activities like bio-control activities, ligninolytic activity, phosphate-solubilisation, production of lytic enzymes, volatile and diffusible antimicrobials, and ammonia production. The presence of such fungal taxa in shifting agricultural fields possibly shows that they can tolerate the stressful conditions such as the high temperature burning operations prior to cultivation. Another crucial property among the beneficial microbes is the solubilisation of phosphorus (P). Much soil P is present in inorganic form that remains insoluble and is not directly available for plants. Most of the bacterial species associated with the rhizosphere can mineralise the inorganic P in the soil and therefore, along with other PGPR traits, such strains could act as effective bio-fertilisers to enhance the growth of crops under shifting cultivation [9].

4 Approaches for Enhancing Soil Health and Fertility Under Shifting Cultivation

Soil fertility under shifting cultivation is determined by physical (e.g. soil texture and soil moisture) and chemical properties (e.g. pH, organic carbon (C), total nitrogen (N), available P, potassium, etc.). It also depends on the biological properties like microbial biomass, composition, and enzyme activities during fallows and under different soil amendments [50]. Microbes in the soil cycle nutrients and regulate C sequestration and emissions. To maintain a continuous supply of nutrients, the presence of microbes in the rhizosphere is a necessity. Deterioration of soil health is of concern for plant health. In this context, soil enzyme activities are important indicators of soil microbial community structure because of their role in the decomposition of soil organic matter and nutrient cycling [51]. Studies that focus on the physical and chemical properties of soil are equally important in evaluating soil health. Various researchers use different parameters like the concentrations of organic C, N, available P, and other micronutrients as indicators of soil health. Based on the quality of the soil, trials with various soil amendments are being undertaken to assess their impacts on soil health and fertility in the North-eastern region of India.

5 Microbe-Based Interventions to Enhance Soil Fertility in Shifting Cultivation Fields

To enhance soil fertility, various alternative approaches are being practiced by shifting cultivation farmers. In general, inputs in the form of good-quality seeds, plant protection chemicals, and plant nutrients like fertilisers and manures are being practiced among some of the farmers [52]. In recent years, to increase crop productivity, the use of chemical fertilisers and pesticides have widely been practiced in agricultural systems [53]. However, productivity also depends on other factors such as climatic conditions. In this context, a multi-cropping system is also practiced under shifting cultivation; however, this is not particularly effective in enhancing the soil fertility or improving the income of the farmers [54]. Therefore, other alternatives such as the use of bio-inoculants and chemical fertilisers are being trialled. Recently, Sharma et al. [55] attempted to analyse the impact of such bioinoculants in combination with different doses of fertiliser on plant health and other criteria like soil N, resident bacterial community, genes, and transcripts involved in the N cycle. Their study demonstrated that the combined use of bioinoculants and 75% of the recommended dose of fertiliser enhanced grain yield of *Cajanus cajan* (Fabaceae) up to 1.28-fold. In addition, such a combination was also found effective for the improvement in soil fertility and the bacterial community involved in the N cycle. Moreover, it also led to better N substrate utilisation with an increase in the transcripts of the *nifH* gene. It was equally discovered to have a positive effect on the soil bacterial community as a more beneficial community with increasing soil N turnover has been observed after its application. Wapongnungsang and Tripathi [50] studied similar effects under shifting cultivation fields in Mizoram. They trailed experimental manipulations, namely microbial inoculation of beneficial rhizosphere microbes (Tmicro+), topsoil (Tsoil+) and litter (Tlitter+) from adjoining forest. The authors observed a notable increase in the soil nutrient status with the addition of Tlitter+. They also observed that soil properties like microbial biomass and enzyme activities in the soil were significantly greater in Tlitter+ followed by Tmicro+ samples. Therefore, this is evidence that the application of Tlitter+ and Tmicro+ to fallows increases the soil fertility and can be considered for subsequent application in shifting cultivation [50]. Furthermore, the response to applications of indigenous soil microbes and rock phosphate on soil fertility and crop productivity in Mizoram was studied [45]. This study showed that a mixture of rock phosphate and microbial cultures from the rhizosphere soil of early regenerating plants was most effective in increasing soil fertility and crop productivity [45].

In shifting cultivation, the changes in the fallow period can influence microbial biomass [56]. Further, modern agricultural practices such as chemical fertiliser and pesticides also have an impact on microbial diversity [57]. The chemicals applied persist in the soil where they can influence a range of soil functions and properties like enzyme activities [58]. As a consequence, examining the impact of both natural and human-managed activities on soil microbial diversity is important for sustainable agricultural production. With the recent advances in molecular methods and high throughput techniques like metagenomics, such studies are gaining attention. Manpoong et al. [10] showed the altered soil biochemical and microbial attributes associated with land-use practices in North-east India are intertwined with changing soil bacterial community structure. They analysed microbial diversity of the different land-uses using high throughput sequencing techniques showing that soil pH was the strongest predictor explaining 96% of the variability in microbial diversity. Acidic soils revealed the presence of acidophilic microbial communities consisting of phyla like Acidobacteria, Pseudomonadota. and Verrucomicrobiota. In addition, organic C content and total N of the soil were also found to have an impact on microbial diversity. Therefore, studies reporting soil amendments [45, 50] that have the potential to influence soil microorganisms and microbial biomass are important for the sustainable shifting agricultural system and need more research attention in the future.

6 Integrated Approaches for Sustainable Shifting Cultivation

Shifting cultivation has constituted an integral part of the indigenous practices across various states of North-east India. Therefore, introducing major changes to this system has been opposed by traditional farming communities, ecologists, and social scientists. Current shifting agricultural practices can be transformed with the preferences of the local farmers without causing any harm to the environment. It is therefore suggested that shifting cultivation may be improved by introducing inputs of seed, fertilisers and others in a phased manner. Moreover, integrated farming with combinations of compatible and region-specific components of agriculture to maintain the productivity of the hill ecosystem with minimal degradation of natural resources can also be performed. However, to date, research on shifting cultivation has largely been limited to management of fallow lands after cultivation. Therefore, extensive research for progressively improving existing shifting cultivation techniques by adapting sustainable integrated farming practices needs more attention [59]. Among the integrated approaches, the use of organic manures combined with beneficial indigenous and adapted microbes is being suggested as the most effective method to maintain healthy and sustainable soil. The application of such methods has demonstrated good promises for increasing crop yields [60]. Most of these methods are based on technological aspects that need specialist skills, therefore, local people involved in shifting cultivation need to be updated with the knowledge and skills of such techniques [53].

Another ecologically-based approach is agroforestry that involves a natural resource management system that can sustain production by integrating trees on farms and in the agricultural landscape. Additionally, in recent years, organic farming is also gaining more research attention and is popular for the production of quality crops. Therefore, long-term organic farming can also reduce the use of synthetic fertilisers and pesticides. Importantly, such aspects need more information about the soil microbial communities. Therefore, understanding microbial diversity and its functioning are of importance before applying these approaches in the fields of shifting cultivation. In this context, microbial collections, that allow us to understand their properties such as beneficial effects on plant growth for sustainable agriculture are important. In the future, successful manipulation of long-term and persistent plant beneficial microbial communities in croplands will benefit current agricultural practices. Based on the current status of knowledge and available technologies, several novel approaches offer the potential to improve the application of beneficial microbial communities in agriculture and thus increase crop productivity under shifting cultivation whilst maintaining environmental sustainability.

Acknowledgements The authors are thankful to the Department of Science and Technology (DST), New Delhi, Government of India for financial support. The work was carried out under financial support from a joint international grant from the Department of Science and Technology (DST) of the Government of India (Project number C/1756/IFD/2019-20) and the Russian Science Foundation (Project number 19-46-02004). The authors would like to extend their sincere appreciation to the Vice Chancellor of Mizoram University, Mizoram, India.

References

1. Bhuyan R (2019) A review note on shifting cultivation in Northeast India amidst changing perceptions. Dhaulagiri J Sociol Anthropol 13:90–95. https://doi.org/10.3126/dsaj.v13i0.24252
2. Thong P, Sahoo UK, Thangjam U, Pebam R (2020) Pattern of forest recovery and carbon stock following shifting cultivation in Manipur, North-East India. PloS One 15:e0239906. https://doi.org/10.1371/journal.pone.0239906
3. Prakash N, Roy SS, Ansari MA, Sharma SK, Punitha P, Sailo B, Singh IM (eds) (2017) Jhum improvement for sustaining farm livelihood and natural resource conservation in North Eastern hill region: vistas and frontiers. ICAR Research Complex for NEH Region, Umiam, Meghalaya, India
4. Bhagawati K, Bhagawati G, Das R, Bhagawati R, Ngachan SV (2015) The structure of *jhum* (traditional shifting cultivation system): prospect or threat to climate. Int Lett Nat Sci 46:16–30. https://doi.org/10.18052/www.scipress.com/ILNS.46.16
5. Swami S (2018) Shifting cultivation: a tribal way of life in north east India and alternative approaches for increasing productivity. Pharma Innov J 7:380–385
6. Ramakrishnan PS, Toky OP (1981) Soil nutrient status of hill agro-ecosystems and recovery pattern after slash and burn agriculture (Jhum) in north-eastern India. Plant Soil 60:41–64. https://doi.org/10.1007/BF02377111
7. Datta J, Gangadharappa NR, Debnath A (2014) Sustainability of *jhum* cultivation as perceived by the tribal people of Tripura. Int J Soc Sci 3:179–190. https://doi.org/10.5958/2321-5771.2014.00100.8
8. Trivedi P, Mattupalli C, Eversole K, Leach JE (2021) Enabling sustainable agriculture through understanding and enhancement of microbiomes. New Phytol 230:2129–2147. https://doi.org/10.1111/nph.17319
9. Banerjee A, Bareh DA, Joshi SR (2017) Native microorganisms as potent bioinoculants for plant growth promotion in shifting agriculture (*jhum*) systems. J Soil Sci Plant Nutr 17:127–140. https://doi.org/10.4067/S0718-95162017005000010
10. Manpoong C, De Mandal S, Bangaruswamy DK, Perumal RC, Benny J, Beena PS, Ghosh A, Kumar NS, Tripathi SK (2020) Linking rhizosphere soil biochemical and microbial community characteristics across different land use systems in mountainous region in Northeast India. Meta Gene 23:100625. https://doi.org/10.1016/j.mgene.2019.100625
11. Philippot L, Raaijmakers JM, Lemanceau P, van der Putten WH (2013) Going back to the roots: the microbial ecology of the rhizosphere. Nat Rev Microbiol 11:789–799. https://doi.org/10.1038/nrmicro3109
12. Song W, Tong X, Zhang J, Meng P, Li J (2017) Autotrophic and heterotrophic components of soil respiration caused by rhizosphere priming effects in a plantation. Plant Soil Environ 63:295–299. https://doi.org/10.17221/233/2017-PSE
13. Thompson LR et al (2017) A communal catalogue reveals Earth's multiscale microbial diversity. Nature 551:457–463. https://doi.org/10.1038/nature24621
14. Pandey A, Chaudhry S, Sharma A, Choudhary VS, Malviya MK, Chamoli S, Rinu K, Trivedi P, Palni LMS (2011) Recovery of *Bacillus* and *Pseudomonas* spp. from the 'fired plots' under shifting cultivation in Northeast India. Curr Microbiol 62:273–280. https://doi.org/10.1007/s00284-010-9702-6
15. Malviya MK, Pandey A, Sharma A, Tiwari SC (2013) Characterization and identification of actinomycetes isolated from 'fired plots' under shifting cultivation in northeast Himalaya, India. Ann Microbiol 63:561–569. https://doi.org/10.1007/s13213-012-0504-x
16. Hauchhum R, Tripathi SK (2020) Impact of rhizosphere microbes of three early colonizing annual plants on improving soil fertility during vegetation establishment under different fallow periods following shifting cultivation. Agric Res 9:213–221. https://doi.org/10.1007/s40003-019-00422-w

17. Griffiths BS, Hallett PD, Kuan HL, Gregory AS, Watts CW, Whitmore AP (2008) Functional resilience of soil microbial communities depends on both soil structure and microbial community composition. Biol Fertil Soil 44:745–754. https://doi.org/10.1007/s00374-007-0257-z
18. Buée M, De Boer W, Martin F, Van Overbeek L, Jurkevitch E (2009) The rhizosphere zoo: an overview of plant-associated communities of microorganisms, including phages, bacteria, archaea, and fungi, and of some of their structuring factors. Plant Soil 321:189–212. https://doi.org/10.1007/s11104-009-9991-3
19. Burns RG, DeForest JL, Marxsen J, Sinsabaugh RL, Stromberger ME, Wallenstein MD, Weintraub MN, Zoppini A (2013) Soil enzymes in a changing environment: current knowledge and future directions. Soil Biol Biochem 58:216–234. https://doi.org/10.1016/j.soilbio.2012.11.009
20. Pant P, Pant P (2017) Ecological restoration techniques for management of degraded, mined-out areas and the role played by rhizospheric microbial communities. In: Singh R, Kumar S (eds) Green technologies and environmental sustainability. Springer, Cham, pp 437–453. https://doi.org/10.1007/978-3-319-50654-8_19
21. Qiu Z, Egidi E, Liu H, Kaur S, Singh BK (2019) New frontiers in agriculture productivity: optimised microbial inoculants and in situ microbiome engineering. Biotechnol Adv 37:107371. https://doi.org/10.1016/j.biotechadv.2019.03.010
22. Baishya LK, Ansari MA, Sarkar GM, Kumar S, Prakash N (2016) Productivity enhancement in shifting cultivated lands through biofertilizer and micro-dosing of NPK in Eastern Himalayan region. Res Crops 17:268–175. https://doi.org/10.5958/2348-7542.2016.00046.2
23. Deka J, Thakuria D, Khyllep A, Ahmed G (2019) Isolation and functional characterization of beneficial bacteria associated with roots of *Thysanolaena maxima* and rhizospheric soil enzymatic activities in jhum agriculture. Curr Agric Res J 7:189–200. https://doi.org/10.12944/CARJ.7.2.07
24. Devi SI, Somkuwar B, Potshangbam M, Talukdar NC (2012) Genetic characterization of *Burkholderia cepacia* strain from North-east India: a potential bio-control agent. Adv Biosci Biotechnol 3:1179–1188. https://doi.org/10.4236/abb.2012.38144
25. Giri K (2019) The first report of indigenous free-living diazotroph *Kosakonia sacchari* isolated from Himalayan alder-based shifting cultivation system in Nagaland, India. J Soil Sci Plant Nutr 19:574–579. https://doi.org/10.1007/s42729-019-00056-5
26. Giri K, Mishra G, Suyal DC, Kumar N, Doley B, Das N, Baruah RC, Bhattacharyya R, Bora N (2023) Performance evaluation of native plant growth-promoting rhizobacteria for paddy yield enhancement in the jhum fields of Mokokchung, Nagaland, North East India. Heliyon 9:e14588. https://doi.org/10.1016/j.heliyon.2023.e14588
27. Hauchhum R, Tripathi SK (2017) Rhizosphere effects of *Melocanna baccifera* on soil microbial properties under different fallow phases following shifting cultivation. Int J Plant Soil Sci 17:34493. https://doi.org/10.9734/IJPSS/2017/34493
28. Hnamte R, Lalruatsanga H, Lalfakzuala R (2021) Isolation, identification and characterization of phosphate solubilizing bacteria from jhum field of Mizoram, India. Geomicrobiol J 38:924–933. https://doi.org/10.1080/01490451.2021.1980922
29. Jain R, Chaudhary D, Dhakar K, Pandey A (2016) A consortium of beneficial fungi survives the fire operations under shifting cultivation in Northeast Himalaya, India. Natl Acad Sci Lett 39:343–346. https://doi.org/10.1007/s40009-016-0482-9
30. Kalita M, Bharadwaz M, Dey T, Gogoi K, Dowarah P, Unni BG, Ozah D, Saikia I (2015) Developing novel bacterial based bioformulation having PGPR properties for enhanced production of agricultural crops. Indian J Exp Biol 53:56–60
31. Kshetri P, Roy SS, Sharma SK, Singh TS, Ansari MA, Sailo B, Singh S, Prakash N (2018) Feather degrading, phytostimulating, and biocontrol potential of native actinobacteria from Northeastern Indian Himalayan region. J Basic Microbiol 58:730–738. https://doi.org/10.1002/jobm.201800169

32. Lyngwi NA, Nongkhlaw M, Kalita D, Joshi SR (2016) Bioprospecting of plant growth promoting Bacilli and related genera prevalent in soils of pristine sacred groves: biochemical and molecular approach. PLoS One 11:e0152951. https://doi.org/10.1371/journal.pone.0152951
33. Majumder M, Shukla AK, Arunachalam A (2008) Nutrient release and fungal succession during decomposition of weed residues in a shifting cultivation system. Commun Biom Crop Sci 3:45–59
34. Momin MD (2000) Characterization of rhizospheric actinomycetes from major crop plants and their plant growth promoting properties under *jhum* fields of Mizoram. PhD thesis, Mizoram University, Aizawl, Mizoram, India. https://repository.tribal.gov.in/handle/123456789/74725
35. Momin MD, Ibrahim KS (2021) Endophytic bacteria isolated from *Ageratum conyzoides* L. and *Mikania micrantha* Kunth of Mizoram, Northeast India. Natl Acad Sci Lett 44:355–358. https://doi.org/10.1007/s40009-020-01013-y
36. Momin MD, Tripathi SK (2018) Rhizospheric actinomycetes from major crop plants under shifting cultivation of Mizoram, Northeast India. In: De D, Roy S, Bera GC (eds) Biotechnology and nature. Kabitika, Medinipur, West Bengal, India, pp 68–75
37. Momin MD, Singh NS, Kumar A, Tripathi SK (2021) Structural and functional characterization of rhizosphere actinomycetes of major crop plants under shifting cultivation practice in Northeast India. Vegetos 34:638–646. https://doi.org/10.1007/s42535-021-00221-1
38. Sangma CBK, Thakuria D (2019) Isolation and screening of cellulose degrading microorganisms from forest floor litters of *jhum* fallows. Proc Natl Acad Sci India Sect B Biol Sci 89:999–1006. https://doi.org/10.1007/s40011-018-1015-8
39. Sarmah R (2010) Arbuscular mycorrhizal fungi in lands abandoned due to shifting cultivation in Karbi Anglong District of Assam (India). http://14.139.116.22:8080/jspui/handle/123456789/133. Accessed 29 Sept 2021
40. Sentimenla (2020) Response of soybean (*Glycine max* L. Merril) to biofertilizer and FYM in the jhum fields under Zunheboto districts of Nagaland, India. Int J Curr Microbiol Appl Sci 9:126–130. https://doi.org/10.20546/ijcmas.2020.906.016
41. Sharmah D, Jha DK (2014) Diversity of arbuscular mycorrhizal fungi in disturbed and undisturbed forests of Karbi Anglong Hills of Assam, India. Agric Res 3:229–238. https://doi.org/10.1007/s40003-014-0110-1
42. Singh NS, Tripathi SK (2020) Temporal changes in soil microbial diversity of tropical and sub-tropical forests of Mizoram, Northeast India. Environ Ecol 38:31–37
43. Singh SS, Tiwari S, Dkhar MS (2003) Species diversity of vesicular-arbuscular mycorrhizal (VAM) fungi in jhum fallow and natural forest soils of Arunachal Pradesh, north eastern India. Trop Ecol 44:205–213
44. Surendirakumar K, Pandey RR, Muthukumar T (2021) Arbuscular mycorrhizal fungi in roots and rhizosphere of black rice in terrace fields of North-East India. Proc Natl Acad Sci India Sect B Biol Sci 91:277–287. https://doi.org/10.1007/s40011-020-01221-y
45. Wapongnungsang, Ovung, EY, Upadhyay KK, Tripathi SK (2021) Soil fertility and rice productivity in shifting cultivation: impact of fallow lengths and soil amendments in Lengpui, Mizoram northeast India. Heliyon 7:e06834. https://doi.org/10.1016/j.heliyon.2021.e06834
46. Mehetre G, Leo VV, Singh G, Dhawre P, Maksimov I, Yadav M, Upadhyaya K, Singh BP (2021) Biocontrol potential and applications of extremophiles for sustainable agriculture. In: Yadav AN, Rastegari AA, Yadav N (eds) Microbiomes of extreme environments. CRC Press, Boca Raton, Florida, USA, pp 230–242
47. Kour D, Rana KL, Yadav N, Yadav AN, Kumar A, Meena VS, Singh B, Chauhan VS, Dhaliwal HS, Saxena AK (2019) Rhizospheric microbiomes: biodiversity, mechanisms of plant growth promotion, and biotechnological applications for sustainable agriculture. In: Kumar A, Meena V (eds) Plant growth promoting rhizobacteria for agricultural sustainability. Springer, Singapore, pp 19–65. https://doi.org/10.1007/978-981-13-7553-8_2
48. Maksimov IV, Sorokan AV, Burkhanova GF, Veselova SV, Alekseev VY, Shein MY, Avalbaev AM, Dhaware PD, Mehetre GT, Singh BP, Khairullin RM (2019) Mechanisms of plant tolerance to RNA viruses induced by plant-growth-promoting microorganisms. Plants 8:575. https://doi.org/10.3390/plants8120575

49. Rillig MC, Mummey DL (2006) Mycorrhizas and soil structure. New Phytol 171:41–53. https://doi.org/10.1111/j.1469-8137.2006.01750.x
50. Wapongnungsang, Tripathi SK (2020) Impact of low cost indigenous soil inputs on soil fertility in different fallow lands following shifting cultivation in Muallungthu, Mizoram. J Indian Soc Soil Sci 68:210–220
51. Datt N, Singh D (2019) Enzymes in relation to soil biological properties and sustainability. In: Meena RS, Kumar S, Bohra JS, Jat ML (eds) Sustainable management of soil and environment. Springer, Singapore, pp 383–406. https://doi.org/10.1007/978-981-13-8832-3_11
52. Upadhaya K, Barik SK, Kharbhih VM, Nongbri G, Debnath G, Gupta A, Ojha A (2020) Traditional bun shifting cultivation practice in Meghalaya, Northeast India. Energy Ecol Environ 5:34–46. https://doi.org/10.1007/s40974-019-00144-3
53. Kumar R, Saurabh K, Kumawat N, Sundaram PK, Mishra JS, Singh DK, Hans H, Krishna B, Bhatt BP (2021) Sustaining productivity through integrated use of microbes in agriculture. In: Seneviratne G, Zavahir JS (eds) Role of microbial communities for sustainability. Springer, Singapore, pp 109–145. https://doi.org/10.1007/978-981-15-9912-5_5
54. Reimeingam M (2017) Shifting cultivation in Manipur: land, labour and environment. J Rural Dev 36:97–119. https://doi.org/10.25175/jrd/2017/v36/i1/112705
55. Sharma R, Shrivas VL, Sharma S (2021) Effect of substitution of chemical fertilizer by bioinoculants on plant performance and rhizospheric bacterial community: case study with *Cajanus cajan*. Braz J Microbiol 52:373–386. https://doi.org/10.1007/s42770-020-00418-7
56. Binarani RK, Yadava PS (2010) Effect of shifting cultivation on soil microbial biomass C, N and P under the shifting cultivations systems of Kangchup Hills, Manipur, North-East India. J Exp Sci 1:14–17
57. Nielsen UN, Osler GHR, Campbell CD, Burslem DFRP, van der Wal R (2010) The influence of vegetation type, soil properties and precipitation on the composition of soil mite and microbial communities at the landscape scale. J Biogeogr 37:1317–1328. https://doi.org/10.1111/j.1365-2699.2010.02281.x
58. Prashar P, Shah S (2016) Impact of fertilizers and pesticides on soil microflora in agriculture. In: Lichtfouse E (ed) Sustainable agriculture reviews, vol 19. Springer, Cham, pp 331–361. https://doi.org/10.1007/978-3-319-26777-7_8
59. Chatterjee D, Kuotsu R, Ray SK, Patra MK, Thirugnanavel A, Kumar R, Borah TR, Chowdhury P, Pongen I, Satapathy BS, Deka BC (2022) Preventing soil degradation in shifting cultivation using integrated farming system models. Arch Agron Soil Sci 68:1841–1857. https://doi.org/10.1080/03650340.2021.1937139
60. Selim MM (2020) Introduction to the integrated nutrient management strategies and their contribution to yield and soil properties. Int J Agron 2821678. https://doi.org/10.1155/2020/2821678

Systematic Review on Shifting Cultivation in North-East India: Emerging Research Directions

Nirmal Debnath, Sabina Yasmin Laskar, Arun Jyoti Nath, Rajiv Pandey, and Ashesh Kumar Das

Abstract Shifting cultivation, or 'slash-and-burn' agriculture, is a age-old agricultural system practiced in South and South-east Asia, including North-east India, using elementary tools and techniques for subsistence crop production. Shifting cultivation was an environmentally sustainable hill agricultural production system, but is currently under threat due to various anthropogenic and natural disturbances. Within the last decade, the sustainability of the practice is under question with the dramatic shortening of the fallow cycle. A suite of factors such as increased pressure to feed the burgeoning population and seizure of public lands, including forests, has compelled shifting cultivators to reduce the fallow period. Therefore, detailed scrutiny of possible barriers and prevailing current mechanisms is needed at cross-sectional levels to resolve the current predicament regarding shifting cultivation practices. This analysis will assist in identifying knowledge gaps to design and guide future research aiming for the sustainability of this socio-ecological system. A systematic literature review of peer-reviewed scientific articles focusing on various domains of shifting cultivation practices in North-east India between 1940 and 2020 was made to synthesise the existing knowledge about the practice. The review identified the uniqueness of mixed cropping in shifting cultivation in North-east India. Arunachal Pradesh contains the greatest diversity of tribal communities practicing shifting cultivation amongst the eight North-east India states. Fallow periods ranged from 4 to 18 years with the longest fallow periods in the state of Meghalaya and the shortest in Assam and Tripura. However, insufficient research has been conducted on the sustainable intensification of shifting cultivation practices thereby restricting the knowledge domain about the existence of the practice in the future. The literature synthesis recommends that future research be reoriented on assessing forest regeneration in shifting cultivation fallow lands and understanding the current socio-ecological linkages within shifting cultivation under global climate change phenomena. The social dimensions need to be evaluated for the sustainability of

N. Debnath · S. Y. Laskar · A. J. Nath (✉) · A. K. Das
Department of Ecology and Environmental Sciences, Assam University, Silchar, Assam 788011, India
e-mail: arunjyotinath@gmail.com

R. Pandey
Indian Council of Forestry Research and Education, Dehradun, Uttarakhand 428006, India

S. K. Tripathi and F. Q. Brearley (eds.), *Shifting Cultivation Systems*,
https://doi.org/10.1007/978-3-031-70388-1_15

shifting cultivation practices for biodiversity conservation, land management, and sustainable livelihoods.

Keywords Livelihoods · Slash-and-burn agriculture · Socio-ecological system · Sustainability · Swidden agriculture

1 Introduction

Shifting cultivation, also referred to as 'swidden' or 'slash-and-burn agriculture' [1], is a unique agricultural system practiced in South and South-east Asia for subsistence crop production [2]. Under shifting cultivation, an area of forest is slashed and burned, cultivated for a few years, and then abandoned for a new area until its fertility has been naturally restored. Currently, shifting cultivation covers roughly 280 million ha across the globe, including newly cultivated fields and all the stages of fallows, with the largest share in Africa, followed by South America and Asia [3, 4]. The practice was an efficient agricultural system in terms of ensuring a subsistence livelihood for indigenous, sparsely distributed communities, and for conserving the ecological integrity of the forested landscape [5–7]. The cultivation cycle in the system is generally composed of three stages: clearing, cropping, and fallowing [8]. Assorted varieties of crops are grown by clearing a patch of vegetation by the slash-and-burn method, using elementary tools and techniques, along with minimal agricultural inputs [9, 10]. This cultivation method is practiced on a rotational basis. The farmers abandon cultivation on the current land and obtain another plot of land for clearing at an interval of every three to five years for crop cultivation [11, 12]. Until the 1970s, these cultivated lands were left uncultivated or abandoned for 15–20 years, known as the 'fallow period', to facilitate vegetation recovery and rejuvenation of soil fertility [2]. The long fallow period was instrumental in improving the soil and establishing the vegetation cover, which made shifting cultivation a sustainable land-use [13, 14]. In recent decades, the increased pressure to feed the burgeoning population and seizure of public lands, including forests, has compelled the shifting cultivators to reduce the fallow period. As a result of these detrimental drivers, the fallow period of the shifting cultivation practice is now reduced to the tune of less than five years from 15 to 20 years in South-east Asia [7, 15–17]. Consequently, short fallowing has now resulted in soil erosion and depletion of soil nutrients [18–25].

North-east India consists of eight states: Arunachal Pradesh, Assam, Manipur, Meghalaya, Mizoram, Nagaland, Tripura, and Sikkim, harbouring more than 100 different tribes that practice shifting cultivation for many generations spread over an area of 1.5 million ha [26, 27]. Shifting cultivation is vital to the economic and cultural identity of the region's rural tribal communities by ensuring livelihood and food security [15]. However, population pressure, urbanisation, and reduced fallow length have adversely influenced this cultivation practice, despite its being a dominant land-use in tropical forested landscapes [3, 28]. Therefore, it is imperative to research various aspects for ensuring efficient productivity of shifting cultivation, such as anthropology and human ecology [15, 29] and on environmental consequences of changing shifting cultivation practices [29, 30]. Given the extensive practice of shifting cultivation in the mountains of the North-east Indian region for decades, investigating the studies on shifting cultivation is vital to understand the status of shifting cultivation in North-east India, along with identification of the knowledge gaps for the sustainability of the practice by designing and guiding future research directions.

2 Methods

2.1 Data Extraction

We performed a systematic review of peer-reviewed scientific articles on shifting cultivation in North-east India using the Web of Science database with permutations of keywords that included 'shifting cultivation', 'jhum cultivation', 'slash and burn cultivation', 'North-east India', 'women and shifting cultivation', 'livelihood and shifting cultivation', 'sustainable intensification', 'fallow management', and 'fallow cycle'. In the first step, the search was focused and applied to the abstract, title and keywords of research articles published between 1940 and June 2020. The literature search identified 804 articles, of which only 319 articles (40%) were exclusively on shifting cultivation in North-east India. In the second step, we excluded the non-peer-reviewed articles published from North-east India, which resulted in selecting a final set of 91 articles (Appendix). These 91 research articles were then evaluated thoroughly.

2.2 Data Analysis and Interpretations

We developed a database based on the information extracted from each article. The database included the year of publication, study location, type of study (review, theoretical, or empirical), name of the tribe practicing shifting cultivation, duration of fallow, crop species, and tree composition. We also recorded the management practices adopted by the farmers.

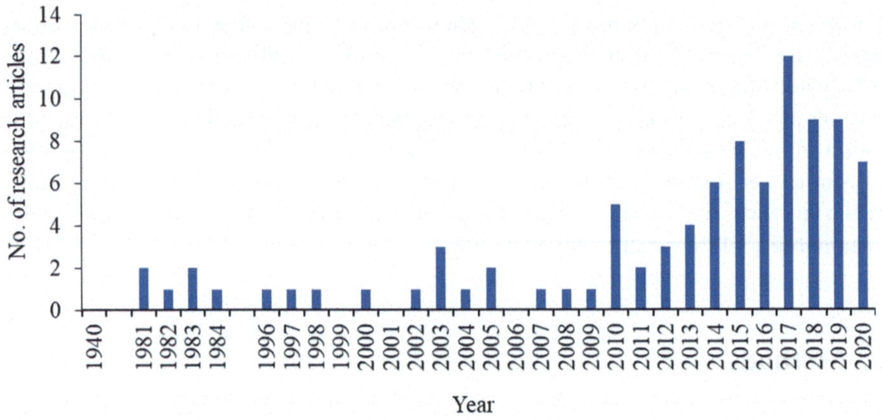

Fig. 1 Temporal distribution of research articles (based on available full-texts) published on shifting cultivation in North-east India between 1940 and 2020

3 Status of Shifting Cultivation in North-East India: An Overview

3.1 Temporal Trends of Shifting Cultivation Studies

The number of research articles on shifting cultivation in North-east India reviewed in this study has rapidly increased since 2010. Two different temporal periods were distinguished, i.e. 1981–2000 and 2000–2020. The purpose of temporal classification was to understand the trend in shifting cultivation research in the last four decades. The analysis reveals that in 20 years, i.e., between 1981 and 2000, only ten articles were published with one article in every two years, in contrast to the subsequent 20 years, i.e. from 2001 to 2020, which accounted for 81 articles with a mean of four articles per year. The maximum number of research articles on the theme was published in 2017 (Fig. 1). Among these studies, 62 articles (68%) were empirical, 20 (22%) were reviews, and 9 (10%) were theoretical.

3.2 Geographical Distribution of Shifting Cultivation Studies

The studies addressing shifting cultivation were geographically unevenly distributed across the states in North-east India (Fig. 2). The greatest number of publications was from studies in Meghalaya (20% of publications; n = 18), and the minimum number was from Assam (2% of publications; n = 2), although in Sikkim there were

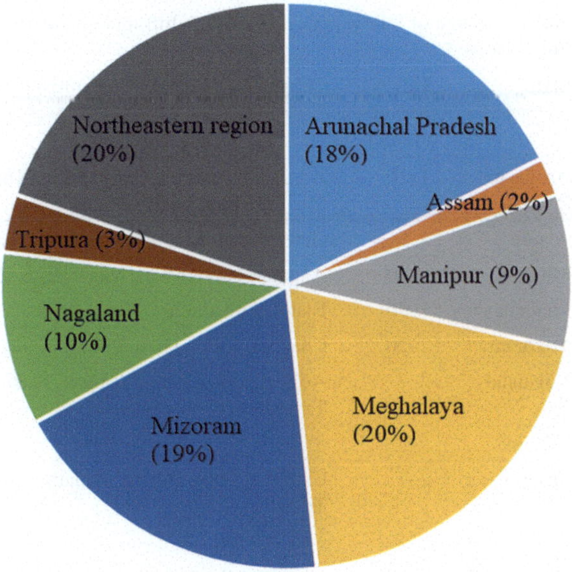

Fig. 2 Geographic distribution of publications on shifting cultivation in North-east India published between 1940 and 2020

no published reports on shifting cultivation (Fig. 2). There were some articles (19%, n = 17) focusing on more than one of the states of North-east India.

3.3 *Fallow Periods and Tribes Involved in Shifting Cultivation*

The focus of articles revolved around the various tribes of North-east India, who practice shifting cultivation. The analysis shows that Arunachal Pradesh harbours the greatest number of tribal communities practicing shifting cultivation with the least in Meghalaya and Mizoram. The different fallow period were reported in the literature for the different states. The lower value in the range indicates the currently practiced shortened fallow period. In contrast, the larger values indicate the historical fallow period being managed in different states in North-east India (Table 1).

3.4 *Cropping Patterns and Crop Constituents of Shifting Cultivation*

The cropping pattern in shifting cultivation in North-east India mainly comprises mixed copping of local/indigenous crop varieties such as staple crops, fruit crops, and grasses. Block plantations (plantations of trees in compact blocks of > 0.1 ha)

Table 1 Range of fallow period length of shifting cultivation studies in North-east India and the tribes involved

State	Fallow period (years)	Tribes practicing shifting cultivation
Arunachal Pradesh	1–10	Adi, Aka, Apatani, Bangni, Idu, Khamba, Memba, Miji, Minyong, Miri, Mishmi, Nishi, Nokte, Padam, Singpho, Tangsa, Wancho
Assam	3–6	Garo, Karbi, Khasi, Kuki, Mizo, Naga
Manipur	5–10	Angami, Kuki, Mizo, Naga, Paite, Zhomi
Meghalaya	1–50	Biate, Garo, Jaintia, Khasi, Karbi
Mizoram	2–20	Chakma, Kuki, Mizo, Pawai, Reang
Nagaland	2–7	Ao, Angami, Chang, Konyak, Kuki, Lotha, Rengma, Sema, Tangkhul, Yimchungur
Sikkim	–	–
Tripura	2–7	Chakma, Garo, Halam, Jamatia, Kuki, Lushai, Mag, Noatia, Reang, Tripuri

is also practiced in Meghalaya (Table 2). Rice (*Oryza sativa*; Poaceae) was the main crop grown in the shifting cultivation fields irrespective of state. A unique traditional practice was reported from Arunachal Pradesh that involves cultivating rice and fish together with millet (*Eleusine coracana*; Poaceae) on the bunds separating agricultural plots [31] (Table 2).

3.5 Management and Conservation Strategies of Shifting Cultivation Areas Adopted by the Cultivators

Diverse management practices under shifting cultivation of North-east India are found in the literature (Table 3). Alder-based (*Alnus nepalensis*; Betulaceae) jhum cultivation and wet terraced rice cultivation are practiced in Nagaland [32] whereas a block pattern of cultivation was prevalent in Meghalaya [33]. In Assam, a select type of shifting cultivation practice is recorded. In this process of cultivation, a patch of forest is cleared completely, the debris is left to dry and then burnt after which the land is used for cultivation for one to two years [33]. After the growing season, the field is abandoned, and weeds and bamboos are allowed to initiate ecological succession until the secondary forest is re-established. In Mizoram, nitrogen-fixing shrubs (*Flemingia macrophylla* and *Tephrosia candida*; both Fabaceae) are planted as dense hedgerows along contour slopes, and a diverse range of crops are cultivated in the inter-row areas [6].

Table 2 Cropping pattern and crops grown in shifting cultivation plots in North-east India

State	Cropping pattern	Crops grown
Arunachal Pradesh	Mixed cropping	Grains and cereals such as maize (*Zea mays*; Poaceae), millet (*Eleusine corocana*; Poaceae), rice (*Oryza sativa*; Poaceae); tuberous crops such as cassava (*Manihot esculenta*; Euphorbiaceae), sweet potato (*Ipomoea batatas*; Convolvulaceae), yam (*Dioscorea* sp.; Dioscoreaceae); vegetable crops such as brinjal (*Solanum melongena*; Solanaceae), chillies (*Capsicum* sp.; Solanaceae), cucumber (*Cucumis sativus*; Cucurbitaceae), gourd (*Luffa acutangula*; Cucurbitaceae), mustard (*Brassica juncea*; Brassicaceae), pumpkin (*Cucurbita maxima*; Cucurbitaceae); fruit such as banana (*Musa* sp.; Musaceae); and spices such as coriander (*Coriandrum sativum*; Apiaceae) and ginger (*Zingiber officinale*; Zingiberaceae)
Assam	Mixed cropping	Upland rice (*Oryza sativa*), vegetables and cash crops such as chillies (*Capsicum* sp.), ginger (*Zingiber officinale*) and sesame (*Sesamum indicum*; Pedaliaceae), etc.
Manipur	Mixed cropping	High-yielding and improved varieties of rice (*Oryza sativa*) along with crops such as air potato (*Dioscorea bulbifera*; Dioscoreaceae), black gram (*Vigna mungo*; Fabaceae), brinjal (*Solanum melongena*), cassava (*Manihot esculenta*), chillies (*Capsicum* sp.), cotton (*Gossypium hirsutum*; Malvaceae), cowpea (*Vigna unguiculata*; Fabaceae), French beans (*Phaseolus vulgaris*; Fabaceae), ginger (*Zingiber officinale*), varieties of maize (*Zea mays*), okra (*Abelmoschus esculentus*; Malvaceae), peas (*Pisum sativum*; Fabaceae), pineapple (*Ananas comosus*; Bromeliaceae), mustard (*Brassica juncea*), soybean (*Glycine max*; Fabaceae), squash (*Sechium edule*; Cucurbitaceae), taro (*Colocasia esculenta*; Araceae), turmeric (*Curcuma longa*; Zingiberaceae) and yams (*Dioscorea* sp.)
Meghalaya	Block plantation; mixed cropping	Millet (*Eleusine corocana*); crops like areca nut (*Areca catechu*; Arecaceae), black pepper (*Piper nigrum*; Piperaceae), broom grass (*Thysanolaena maxima*; Poaceae), cashew nut (*Anacardium occidentale*; Anacardiaceae), cotton (*Gossypium hirsutum*), ginger (*Zingiber officinale*), maize (*Zea mays*), rice (*Oryza sativa*), rubber (*Hevea brasiliensis*; Euphorbiaceae), taro (*Colocasia esculenta*), tea (*Camellia sinensis*; Theaceae) and turmeric (*Curcuma longa*); vegetables like air potato (*Dioscorea bulbifera*), cucumber (*Cucumis sativus*), French beans (*Phaseolus vulgaris*), okra (*Abelmoschus esculentus*), peas (*Pisum sativum*), potato (*Solanum tuberosum*; Solanaceae), radish (*Raphanus raphanistrum* subsp. *sativus*; Brassicaceae), etc.

(continued)

Table 2 (continued)

State	Cropping pattern	Crops grown
Mizoram	Mixed cropping	Rice (*Oryza sativa*) as an intercrop with banana (*Musa* sp.), brinjal (*Solanum melongena*), chillies (*Capsicum* sp.), cowpea (*Vigna unguiculata*; Fabaceae), maize (*Zea mays*), mangoes (*Mangifera indica*; Anacardiaceae), pineapple (*Ananas comosus*) pumpkin (*Cucurbita andreana*; Cucurbitaceae) and sesame (*Sesamum indicum*)
Nagaland	Mixed cropping	Banana (*Musa* sp.), cassava (*Manihot esculenta*), French beans (*Phaseolus vulgaris*), ginger (*Zingiber officinale*), lemon (*Citrus limon*; Rutaceae), maize (*Zea mays*), orange (*Citrus* spp.; Rutaceae), pineapple (*Ananas comosus*), potato (*Solanum tuberosum*), sweet potato (*Ipomoea batatas*), taro (*Colocasia esculenta*) and upland rice (*Oryza sativa*)
Tripura	Mixed cropping	Barley (*Hordeum vulgare*; Poaceae), basil (*Ocimum americanum*; Lamiaceae), brinjal (*Solanum melongena*), cassava (*Manihot esculenta*), chillies (*Capsicum* sp.), culantro (*Eryngium foetidum*; Apiaceae), elephant foot yam (*Amorphophallus paeoniifolius*; Araceae), foxtail millet (*Setaria italica*; Poaceae), ginger (*Zingiber officinale*), maize (*Zea mays*), pigeon pea (*Cajanus cajan*; Fabaceae), rice (*Oryza sativa*), roselle (*Hibiscus sabdariffa*; (Malvaceae), sesame (*Sesamum indicum*), sorghum (*Sorghum bicolor*; Poaceae), taro (*Colocasia esculenta*), turmeric (*Curcuma longa*), wax gourd (*Benincasa hispida*; Cucurbitaceae) and winged bean (*Psophocarpus tetragonolobus*; Fabaceae)

3.6 Role of Women in Managing Shifting Cultivation Farms

The literature reveals the active participation of women in shifting cultivation across the states of North-east India and, in general, women make an essential contribution to the practice [34]. Men's involvement in shifting cultivation begins from selecting land while women's activities commence from slashing and clearing. In most cases, males set the fire on the selected plot, and females control the fire so that it does not go beyond the demarcated areas. Women level the field using spades and then sow the seeds.

3.7 Soil Physical and Chemical Properties

The plot size used for studying vegetation attributes in different shifting cultivation studies in North-east India varied from 0.0015 to 6.9 ha. Most of the studies considered a plot size of 0.01 ha area. The present analysis revealed different soil physical

Table 3 Land-use management and conservation strategies of shifting cultivation areas adopted by cultivators in Northeast India

State	Land-use management and conservation strategies
Arunachal Pradesh	In addition to the slash-and-burn method, the Adi farmers control the weeds, soil-borne pathogens, and other diseases of crops through uprooting the weeds thrice in a season and mulching them on the bund or depositing and burning in the fields. Such ashes (locally known as 'mekor') help in adding nutrients to the soil. On the other hand, the mulched weed biomass assists in minimising the leaching of nutrients by releasing them slowly
Assam	Weeds and bamboos are allowed to grow in the shifting cultivation fields abandoned after one or two growing seasons. This facilitates the re-establishment of the secondary forest for the areas to be re-used for cultivation
Meghalaya	Traditionally in Meghalaya, the shifting cultivation practices are of two types: traditional shifting cultivation and bun cultivation. The buns are usually 2–4 m long, 1–2 m wide, and 0.15–0.35 m in height and spaced at 1–2 m depending on the soil depth. A thin layer of soil covers the biomass before burning under anaerobic conditions to slow biomass conversion into ash
Mizoram	Nitrogen-fixing shrubs such as *Flemingia macrophylla* and *Tephrosia candida* (both Fabaceae) are planted as dense hedgerows along the slope contour, and a diverse range of crops are grown in the inter-row areas. In addition to nitrogen fixation, *Flemingia macrophylla* stems prevent wind damage and erosion of soil. In contrast, *Tephrosia candida* suppresses weed growth in the cultivation plots by producing many seeds and seedlings
Nagaland	The crop nutrient requirements in shifting cultivation fields are fulfilled through the input of the dung of cows, mithun, pigs, and other animals. For harvesting, the Naga farmers cut or pick the economically valuable part of the crops such as rice panicles, maize cobs, etc. and leave the residues at the top of the slopes to reduce soil erosion. This method also assists in enhancing soil fertility. Another unique method observed here was the alder-based jhum cultivation

and chemical properties, including soil texture, bulk density, pH, organic carbon, total nitrogen, available phosphorus, exchangeable potassium, etc. had been studied in the shifting cultivation soils. Clear changes occur in soil macro- (> 250 μm) and microaggregates and porosity under fallow lands because of soil densification under a short fallow period [35]. The soil under short fallow is prone to a high nutrient loss (60–700% more than native forest), resulting in low agronomic yields [26, 36]. Several authors have reported the effect of the fallow period on soil physical and chemical properties compared to the native forest. However, the literature search revealed different studies considered different soil depths ranging from 10 to 90 cm (Fig. 3), leading to uncertainty in comparing the different studies. Among these, the soil depths considered by the majority of the studies were 0–15 cm, 0–20 cm, and 0–30 cm depth (Fig. 3).

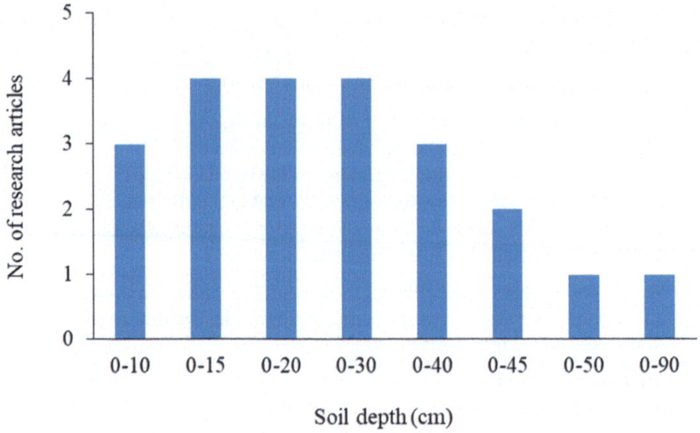

Fig. 3 Different soil depths considered to assess various soil physico-chemical parameters in different studies on shifting cultivation in North-east India

3.8 Shifting Cultivation: Sustainability in Question

The dramatic shortening of the fallow period from 20 to 30 years to three to five years [37] has raised concerns about the sustainability of shifting cultivation. The reduction in the fallow period causes a decline in productivity [35] due to the associated soil erosion and nutrient loss, which ultimately leads to ecosystem disservices [38]. Increasing the frequency of shifting cultivation cycles and shortening the fallow period leads to losses of biodiversity and ecosystem services on a regional as well as local scale [39], and a vicious cycle of poverty and reduced quality of life for populations practicing shifting cultivation [40, 41]. Preparation of agricultural land through slash-and-burn is also a source of greenhouse gas emissions since carbon dioxide (CO_2) produced from biomass burning accumulates in the atmosphere [42, 43]. These systems are usually practiced in tropical environments characterised by heavy rains with high erosivity and structurally weak soils with low-activity clays that are prone to rapid degradation [35, 44]. Additionally, the cultivation practices cause fragmentation of habitats, reductions in native species and invasion by exotic weeds [45]. The cost–benefit analysis shows that the livelihood, food security, and socio-cultural events of the concerned communities are given more importance than the environmental issues associated with it [45].

4 Viable Alternatives to Shifting Cultivation in North-East India

Given the numerous impacts of the current practices of shifting cultivation with a reduced fallow period, it is important to develop viable and sustainable approaches to this practice in the interest of the indigenous communities. Shifting cultivation, a forest-agriculture landscape of the tropics, is rapidly being transformed to other land-uses in South-east Asia with limited knowledge of the consequences of the transitions, thereby creating opportunities and challenges for being incentivised under the reduced greenhouse gas emissions from deforestation and degradation (REDD+) mechanism [46]. Perennialisation of agricultural lands has been suggested to mitigate climate change and enhance food security and ecosystem service delivery. Agroforestry can be an alternative to shifting cultivation with some incentives in North-east India [26] and recommended linking agroforestry systems to REDD+ strategies for the lands with high local-scale expansion and intensification potential of shifting cultivation. In the absence of appropriate incentive-based mechanisms for fallow land management, the dependent population may not be interested in adopting the options due to poor economic rewards. Once the appropriate guidelines are planned, the government may incentivise state agencies to formalise, promote, and organise marketing of such farmers' produce by creating provisions for alternate income-generation sources and opportunities for youth employment [47].

Nath et al. [48] discussed the feasibility of introducing the 'grains for ecosystem carbon management' by providing food grains to the shifting cultivator as an alternative to shifting cultivation. Efforts were also observed to upscale local livelihoods and sustainability on sloping land by incorporating local knowledge with resource-based technologies and nature-based approaches. Among them, agroforestry, watershed-based production units, multilayered home gardens, contour planting, incorporation of horticultural crops, community forest management, and seepage water harvesting are potential technologies adopted by the hill dwellers for sustainable livelihoods [49, 50]. More documentation of local knowledge-based technologies, refinement and dissemination will empower the hill dwellers in managing their lands more sustainably. The sustainable intensification of agroecosystems is needed to transform shifting cultivation systems into intensive use [38]. We propose a sustainable intensification model to restore depleted soil for intensive use (Fig. 4). Thus, the promotion of tree-based crop production models in shifting cultivation areas in North-east India would be a viable option for sustaining the agro-biodiversity, forest cover, and soil carbon storage while improving the socioeconomic status of the farmers in the mountain areas (Fig. 5).

Fig. 4 Proposed model for sustainable intensification of shifting cultivation agricultural land for improving soil health and enhancing farm productivity

5 Future Research Priorities

Future research should be conducted based on the potential of shifting cultivation to improve the well-being of the people practicing it, the challenges and barriers in social and ecological aspects, and the potentiality and linkages to environmental management. Future research may focus on a variety of aspects, such that the co-existence of the indigenous communities and their ecological relationships with their environment should remain intact. Specific research directions may include the following:

- Develop and synthesise the current state of knowledge regarding shifting cultivation practices, production, utilisation, value addition and marketing of shifting cultivation products.
- Evaluation of the current linkages and future opportunities between women and youths with shifting cultivation and developing viable business enterprises for the shifting cultivation produce.

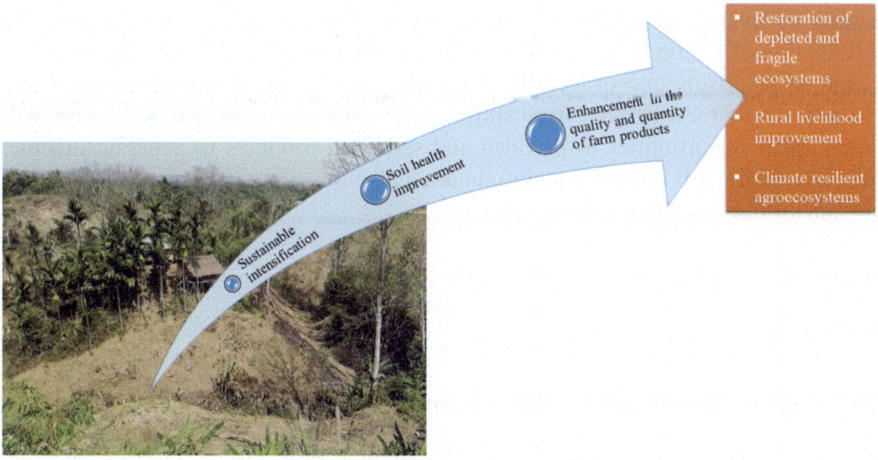

Fig. 5 Schematic presentation of sustainable tree-based crop production system for food security and nature conservancy in shifting cultivation landscapes of North-east India

- Identifying germplasm for higher yield and protein content, farm diversification, income, and livelihoods for hill farmers practicing shifting cultivation along with adoption mechanism by these cultivators.
- Development of community-based sustainable agricultural intensification systems to make shifting cultivation a fully sustainable agricultural practice that can promote the ecological integrity of degraded forests while minimising the economic risks of current practices.
- Development of predictive models to assess the impact of land-use and farming systems on soil quality, water balance, microclimate, agronomic productivity, and sustainability.
- Cost–benefit analysis of practising shifting cultivation vis-a-vis other agroecological systems.
- Exploring and developing protocols for incentive-based mechanisms for strengthening shifting cultivation such as the REDD+ mechanism leading to facilitation for mitigation and adaptation to climate change.
- Evaluation of adoption barriers and challenges of shifting cultivation in relation to the indigenous communities along with evaluation of cultural aspects of the practice.

6 Conclusions

The present chapter provides insights into the current research status on shifting cultivation in North-east India and future research needs. We suggest developing a consistent and standard protocol to study vegetation and soil attributes for future research to generalise research outputs. Given the sustainability concerns of current shifting cultivation practices and its significant impacts on agro-ecosystems, the introduction of viable alternatives to shifting cultivation, for instance, agroforestry, should be positively encouraged. If an enabling policy environment is created and demand for products derived from shifting cultivated farms increases, farmers will have market opportunities to produce and sell such products to generate income. Such a strategy will conserve and maintain agrobiodiversity while advancing food and nutritional security.

Appendix

Papers used in the analysis of shifting cultivation trends in North-east India

Reference	Journal
Arunachalam (2002) Dynamics of soil nutrients and microbial biomass during first year cropping in an 8-year jhum cycle	Nutrient Cycling in Agroecosystems
Arunachalam and Arunachalam (2002) Evaluation of bamboos in eco-restoration of 'jhum' fallows in Arunachal Pradesh: ground vegetation, soil and microbial biomass	Forest Ecology and Management
Barthakur (2004) Shifting cultivation in North-east region	Artha Journal of Social Sciences
Behera et al. (2016) From jhum to broom: agricultural land-use change and food security implications on the Meghalaya Plateau, India	Ambio
Bhagawati et al. (2015) The structure of jhum (traditional shifting cultivation system): prospect or threat to climate	International Letters of Natural Sciences
Bhatt and Laxminarayana (2010) Restoration of shifting cultivation areas through hedgerow species in eastern Himalaya, India	Journal of Non-Timber Forest Products
Bhatt et al. (2010) Rehabilitation of shifting cultivation areas through agroforestry: a case study in Eastern Himalaya, India	Journal of Tropical Forest Science
Bhuyan and Teyang (2015) Assessment of traditional shifting cultivation practiced by ethnic Nocte and Wancho of eastern Himalaya	International Journal of Modern Science and Engineering Technology
Bos et al. (2020) Exploring trajectories of shifting-cultivation landscapes through games the case of Assam India	Sustainability
Chase and Singh (2014) Soil nutrients and fertility in three traditional land use systems of Khonoma, Nagaland, India	Resources and Environment

(continued)

(continued)

Reference	Journal
Chaudhury and Sarkar (2020) Shifting cultivation in North-eastern India: a comparative analysis	Studies in Indian Place Names
Das and Das (2014) Shifting cultivation in Tripura—a critical analysis	Journal of Agriculture and Life Sciences
Das and Saha (2013) Study of the shifting cultivation scenario in Aizwal district, Mizoram using remote sensing techniques	Journal of Soil and Water Conservation
Deb et al. (2013) Technological innovations in shifting agricultural practices by three tribal farming communities of Meghalaya, northeast India	Tropical Ecology
Deka and Mishra (1982) Decomposition of bamboo (*Dendrocalamus hamiltonii* Nees.) leaf-litter in relation to age of jhum fallows in Northeast India	Plant and Soil
de Maaker et al. (2016) Shifting ground? State and market in the uplands of Northeast India	The Newsletter
Devi and Choudhury (2013) Soil fertility status in relation to fallow cycles and landuse practices in shifting cultivated areas of Chandel district Manipur, India	IOSR Journal of Agriculture and Veterinary Science
Giri K (2019) The first report of indigenous free-living diazotroph *Kosakonia sacchari* isolated from Himalayan alder-based shifting cultivation system in Nagaland, India	Journal of Soil Science and Plant Nutrition
Gogoi et al. (2020) Vegetation and ecosystem carbon recovery following shifting cultivation in Mizoram-Manipur-Kachin rainforest eco-region, Southern Asia	Ecological Processes
Goswami et al. (2012) Factors influencing farmers' adoption of slash and burn agriculture in North East India	Forest Policy and Economics
Hauchhum and Tripathi (2017) Rhizosphere effects of *Melocanna baccifera* on soil microbial properties under different fallow phases following shifting cultivation	International Journal of Plant and Soil Science
Jha and Lalnunmawia (2003) Agroforestry with bamboo and ginger to rehabilitate degraded areas in North East India	Journal of Bamboo and Rattan
Kumar et al. (2010) Designing a protected area network for conservation planning in *Jhum* landscapes of Garo Hills, Meghalaya	Journal of the Indian Society of Remote Sensing

(continued)

(continued)

Reference	Journal
Kumar et al. (2017) Validation of common salt application on productivity, profitability, nutrient uptake and soil health of upland rice (*Oryza sativa* L.) under shifting cultivation area of Nagaland	Indian Journal of Traditional Knowledge
Kuotsuo et al. (2014) Shifting cultivation: an 'organic like' farming in Nagaland	Indian Journal of Hill Farming
Kurien et al. (2019) Farms or forests? Understanding and mapping shifting cultivation using the case study of West Garo Hills, India	Land
Lhungdim et al. (2013) Upgrading shifting rice cultivation for ecological stability in North Eastern region through nutrient supplements	Environment and Ecology
Lungmuanaa et al. (2017) Impact of secondary forest fallow period on soil microbial biomass carbon and enzyme activity dynamics under shifting cultivation in North Eastern Hill region, India	Catena
Lungmuana et al. (2018) Impact of post burn jhum agriculture on soil carbon pools in the north-eastern Himalayan region of India	Soil Research
Malviya et al. (2018) Chitinase and glucanase activities of antagonistic Streptomyces spp isolated from fired plots under shifting cultivation in Northeast India	Journal of Advanced Research in Biotechnology
Mishra and Ramakrishnan (1983) Slash and burn agriculture at higher elevations in north-eastern India. II. Soil fertility changes	Agriculture, Ecosystems and Environment
Mishra and Ramakrishnan (1984) Nitrogen budget under rotational bush fallow agriculture (jhum) at higher elevations of Meghalaya in north-eastern India	Plant and Soil
Mishra et al. (2017) Soil quality assessment under shifting cultivation and forests in North Eastern Himalaya of India	Archives of Agronomy and Soil Science
Mishra et al. (2019) Modeling soil organic carbon dynamics under shifting cultivation and forests using RothC model	Ecological Modelling
Pandey et al. (2020) Indigenous people's attachment to shifting cultivation in the Eastern Himalayas, India: a cross-sectional evidence	Forest Policy and Economics
Pasha et al. (2020) Assessment of shifting cultivation fallows in Northeastern India using Landsat imageries	Tropical Ecology

(continued)

(continued)

Reference	Journal
Punitha et al. (2018) Shifting cultivation in North East India: social dimension, cross cultural reflection and strategies for improvement	Indian Journal of Agricultural Sciences
Punitha et al. (2020) *Jhumias* of Manipur in North East India: socio-economic characteristics and *jhum* activities	Indian Research Journal of Extension Education
Rai (2005) Apatani paddy-cum-fish cultivation: An indigenous hill farming system of North East India	Indian Journal of Traditional Knowledge
Ramakrishnan and Toky (1981) Soil nutrient status of hill agro-ecosystems and recovery pattern after slash and burn agriculture (jhum) in north-eastern India	Plant and Soil
Reimeingam (2017) Shifting cultivation in Manipur: land, labour and environment	Journal of Rural Development
Reza et al. (2014) Microbial biomass and enzyme activity in relation to shifting cultivation and horticultural practices in humid subtropical North-Eastern India	Range Management and Agroforestry
Riahtam et al. (2018) Assessment of shifting cultivation dynamics in East Garo Hills District, Meghalaya, India	IOP Conference Series: Earth and Environmental Science
Saha et al. (2011) Soil erodibility characteristics under modified land-use systems as against shifting cultivation in hilly ecosystems of Meghalaya, India	Journal of Sustainable Forestry
Saha et al. (2012) Soil health management under hill agroecosystem of North East India	Applied and Environmental Soil Science
Sahoo (1997) The viable weed seed content of 'jhum' shifting agriculture and terrace agroecosystems of Meghalaya	Seed Research
Saikia and Bhaduri (2012) An institutional analysis of transition to market: The case of shifting cultivators in Mon, Nagaland	International Journal of Rural Management
Saplalrinliana et al. (2016) Impact of shifting cultivation on litter accumulation and properties of *jhum* soils of North East India	Journal of the Indian Society of Soil Science
Sarkar et al. (2015) Potential of fallow chronosequence in shifting cultivation to conserve soil organic carbon in northeast India	Catena
Sati (2019) Shifting cultivation in Mizoram, India: an empirical study of its economic implications	Journal of Mountain Science

(continued)

(continued)

Reference	Journal
Sati and Rinawma (2014) Practices of shifting cultivation and its implications in Mizoram, North-east India: a review of existing research	Nature and Environment
Shankar Raman et al. (1998) Recovery of tropical rainforest avifauna in relation to vegetation succession following shifting cultivation in Mizoram, north-east India	Journal of Applied Ecology
Tawnenga et al. (1996) Evaluating second year cropping on jhum fallows in Mizoram, north-eastern India—phytomass dynamics and primary productivity	Journal of Biosciences
Shimrah et al. (2015) The shifting agricultural system (*jhum*) and strategies for sustainable agroecosystems in Northeast India	Agroecology and Sustainable Food Systems
Shimrah T (2017) Forest, shifting agriculture and strategies for sustainable landscape management in traditional mountain communities of Northeast India	National Academy Science Letters
Sillitoe (2008) Shifting cultivation and sustainable development of North-eastern India: tradition in transition	Mountain Research and Development
Singh (2019) Sustainable alternatives to shifting cultivation in North East India	International Journal of Advanced Scientific Research and Management
Singh et al. (2003) Species diversity of vesicular–arbuscular mycorrhizal (VAM) fungi in jhum fallow and natural forest soils of Arunachal Pradesh, north eastern India	Tropical Ecology
Singh et al. (2014) Land use impact on soil quality in eastern Himalayan region of India	Environmental Monitoring and Assessment
Swami (2018) Shifting cultivation: A tribal way of life in north east India and alternative approaches for increasing productivity	The Pharma Innovation Journal
Tangjang (2009) Traditional slash and burn agriculture as a historic land use practice: a case study from the ethnic Noctes in Arunachal Pradesh, India	World Journal of Agricultural Sciences
Teegalapalli and Datta (2016) Field to a forest: patterns of forest recovery following shifting cultivation in the Eastern Himalaya	Forest Ecology and Management
Teegalapalli and Datta (2016) Shifting to settled cultivation: changing practices among the Adis in central Arunachal Pradesh, north-east India	Ambio

(continued)

(continued)

Reference	Journal
Teegalapalli et al. (2018) Recovery of soil macronutrients following shifting cultivation and ethnopedology of the Adi community in the Eastern Himalaya	Soil Use and Management
Teegalapalli et al. (2019) Factors affecting the survival of tree seedlings following shifting cultivation in the Eastern Himalaya	bioRxiv
Thong et al. (2018) Influence of socio-economic factors on the existing shifting cultivation practice in Champhai district of Mizoram	Journal of Hill Agriculture
Toky and Ramakrishnan (1981) Cropping and yields in agricultural systems of the north-eastern hill region of India	Agro-Ecosystems
Toky and Ramakrishnan (1983) Secondary succession following slash and burn agriculture in North-Eastern India: I. Biomass, litterfall and productivity	Journal of Ecology
Tripathi et al. (2015) Vegetation change during recovery of shifting cultivation (jhum) fallows in a subtropical evergreen forest ecosystem of North-eastern India	International Journal of Plant and Environment
Verma et al. (2017) Alternatives of shifting cultivation in North-eastern region of India	Report and Opinion
Wapongnungsang et al. (2017) Litter decomposition vis-a-vis carbon and nitrogen dynamics of *Tephrosia candida* components in different fallow periods following shifting cultivation in Mizoram	Indian Journal of Ecology
Wapongnungsang et al. (2018) Changes in soil fertility and rice productivity in three consecutive years cropping under different fallow phases following shifting cultivation	International Journal of Plant and Soil Science
Yadav et al. (2012) Impact of slash-and-burn agriculture on forest ecosystem in Garo Hills landscape of Meghalaya, North-East India	Journal of Biodiversity Management and Forestry
Zodinpuii et al. (2019) Earthworm diversity, density and distribution under shifting (jhum) cultivation in a tropical hilly terrain of Mizoram, North East India	Journal of Environmental Biology
Reference	**Book**
Bhatt et al. (2007) Shifting cultivation: issues and strategies in Northeastern hill region	In: Biodiversity and its Significance (eds P Tandon et al), K International, New Delhi, India

(continued)

(continued)

Reference	Journal
Ghosh and Ali (2007) Emerging dimension of shifting cultivation in Arunachal Pradesh	In: Natural Resource Management And Sustainable Development In North-East India (ed. N Ali), Mittal Publications, New Delhi, India
Kant and Katwal (2003) Exploring possibilities of reforestation of forests exposed to shifting cultivation and encroachment in North East India through clean development	In: Proceedings of the National Workshop on Technological Innovations and Research Advancements for Application in Joint Forest Management, ICFRE, New Forest, Dehradun (ed. Anon.), Forest Research Institute, Dehradun, Uttaranchal, India
Kumar et al. (2016) Towards the natural resource management for resilient shifting cultivation system in eastern Himalayas	Conservation Agriculture: an Approach to Combat Climate Change in Indian Himalaya (eds JK Bisht et al), Springer, Singapore
Mipun and Saika (2003) Landuse/landcover detection mapping and identification of shifting cultivation areas of Tirap district	In: Environment, Locational Decisions and Regional Planning (eds S. Singh et al), The Geographical Society of North-Eastern Hill Region, North Eastern Hill University, Shillong, Meghalaya, India
Nath et al. (2016) Soil and jhum cultivation	In: Encyclopedia of Soil, 3rd ed (ed R Lal), CRC Press, Boca Raton, FL, USA
Pandey et al. (2011) *Jatropha curcas*: a solution for moisture stress shifting cultivation areas of Northeast region	In: Biofuels: Potential and Challenges (eds AK Pandey & AK Mandal) Scientific Publishers, Jodhpur, Rajasthan, India
Samal et al. (2019) Local knowledge in managing upland agriculture by the Adis in Arunachal Pradesh, Northeast India	In: Shifting Perspectives in Tribal Studies (ed MC Behera), Springer Nature, Singapore
Sati (2020) Ecological implications of shifting cultivation	In: Economic and Ecological Implications of Shifting Cultivation in Mizoram, India (ed VP Sati), Springer Nature, Switzerland
Tiwari (2003) Innovations in shifting cultivation, land use and land cover change in higher elevations of Meghalaya, India	In: Methodological Issues in Mountain Research. a Socio-Ecological System Approach (eds. PS Ramakrishnan et al), Oxford & IBH Publishing Co., New Delhi, India
Tripathi et al. (2017) Shifting cultivation on steep slopes of Mizoram, India: impact of policy reforms	In: Shifting Cultivation Policies: Balancing Environmental and Social Sustainability (ed. M Cairns), CABI, Wallingford, UK
Reference	**Dissertation**
Lalramliani (2017) Comparative study of impact of shifting cultivation and settled agriculture on soil fertility in Aizawl and Lunglei districts of Mizoram	Doctoral dissertation, Mizoram University
Reference	**Website**

(continued)

(continued)

Reference	Journal
Anon (2004) The Shillong Declaration on Shifting Cultivation in the Eastern Himalayas	https://www.burmalibrary.org/docs20/Shillong_Declaration-2004-10.pdf [Accessed 28 Sept 2020]
Reimeingan (2017) Usage of Land and Labour under Shifting Cultivation in Manipur: Working Paper 384	http://www.isec.ac.in/WP%20384%20-%20Marchang%20Reimeingam_4%20-%20Final.pdf [Accessed 28 Sept 2020]
Sarmah (2010) Arbuscular mycorrhizal fungi in lands abandoned due to shifting cultivation in Karbi Anglong District of Assam (India)	http://dlkkhsou.inflibnet.ac.in/bitstream/123456789/133/1/22.pdf [Accessed 28 Sept 2020]

References

1. Mertz O, Padoch C, Fox J, Cramb RA, Leisz SJ, Lam NT, Vien TD (2009) Swidden change in Southeast Asia: understanding causes and consequences. Hum Ecol 37:259–264. https://doi.org/10.1007/s10745-009-9245-2
2. Rasul G, Thapa GB (2003) Shifting cultivation in the mountains of South and Southeast Asia: regional patterns and factors influencing the change. Land Degrad Dev 14:495–508. https://doi.org/10.1002/ldr.570
3. Heinimann A, Mertz O, Frolking S, Christensen AE, Hurni K, Sedano F, Parsons Chini L, Sahajpal R, Hansen M, Hurtt G (2017) A global view of shifting cultivation: recent, current, and future extent. PLoS One 12:e0184479. https://doi.org/10.1371/journal.pone.0184479
4. Li P, Feng Z, Jiang L, Liao C, Zhang J (2014) A review of swidden agriculture in Southeast Asia. Remote Sens 6:1654–1683. https://doi.org/10.3390/rs6021654
5. Bruun TB, de Neergaard A, Lawrence D, Ziegler AD (2009) Environmental consequences of the demise in swidden cultivation in Southeast Asia: carbon storage and soil quality. Hum Ecol 37:375–388. https://doi.org/10.1007/s10745-009-9257-y
6. Grogan P, Lalnunmawia F, Tripathi SK (2012) Shifting cultivation in steeply sloped regions: a review of management options and research priorities for Mizoram state, Northeast India. Agrofor Syst 84:163–177. https://doi.org/10.1007/s10457-011-9469-1
7. Ramakrishnan PS (1992) Shifting agriculture and sustainable development: an interdisciplinary study from North-eastern India, UNESCO Man and the Biosphere Series, vol 10. UNESCO and Parthenon Publishing Group, Paris, France
8. Kleinman P, Pimentel D, Bryant R (1995) The ecological sustainability of slash-and-burn agriculture. Agric Ecosyst Environ 52:235–249. https://doi.org/10.1016/0167-8809(94)00531-I
9. Geertz C (1963) Agricultural involution: the processes of ecological change in Indonesia. University of California Press, Berkeley, CA, USA
10. Ruthenberg H (1976) Farming systems in the tropics, 2nd edn. Clarendon Press, Oxford, UK
11. Conklin HC (1957) Hanunóo agriculture: a report on an integral system of shifting cultivation in the Philippines. Forestry Development Paper No. 12. FAO, Rome, Italy
12. Spencer JE (1966) Shifting cultivation in Southeastern Asia. University of California Press, Berkeley, CA, USA
13. Ramakrishnan PS (1984) The science behind rotational bush fallow agriculture system (jhum). Proc Indian Acad Sci 93:379–400. https://doi.org/10.1007/BF03053090
14. Warner K (1991) Shifting cultivators: local technical knowledge and natural resource management in the humid tropics. Community Forestry Note 8. FAO, Rome, Italy

15. Bhuyan R (2019) A review note on shifting cultivation in Northeast India amidst changing perceptions. Dhaulagiri J Sociol Anthropol 13:90–95. https://doi.org/10.3126/dsaj.v13i0.24252
16. Eastmond A, Faust B (2006) Farmers, fires, and forests: a green alternative to shifting cultivation for conservation of the Maya forest? Landsc Urban Plan 74:267–284. https://doi.org/10.1016/j.landurbplan.2004.09.007
17. Ziegler AD, Bruun TB, Guardiola-Claramonte M, Giambelluca TW, Lawrence D, Lam NT (2009) Environmental consequences of the demise in Swidden cultivation in montane mainland Southeast Asia: hydrology and geomorphology. Hum Ecol 37:361–373. https://doi.org/10.1007/s10745-009-9258-x
18. Alegre JC, Cassel DK (1996) Dynamics of soil physical properties under alternative systems to slash-and-burn. Agric Ecosyst Environ 58:39–48. https://doi.org/10.1016/0167-8809(95)00654-0
19. Baker R (1984) Protecting the environment against the poor. Ecologist 14:53–60
20. Brady NC (1996) Alternatives to slash-and-burn: a global imperative. Agric Ecosyst Environ 58:3–11. https://doi.org/10.1016/0167-8809(96)00650-0
21. Gafur A, Borggaard OK, Raunsø Jensen J, Petersen L (2000) Changes in soil nutrient content under shifting cultivation in the Chittagong Hill Tracts of Bangladesh. Dan J Geogr 100:37–46. https://doi.org/10.1080/00167223.2000.10649437
22. Gafur A, Raunsø Jensen J, Borggaard OK, Peterson L (2003) Run-off and losses of soil nutrients from small watershed and shifting cultivation (jhum) in the Chittagong Hill Tracts of Bangladesh. J Hydrol 274:30–46. https://doi.org/10.1016/S0022-1694(02)00351-7
23. Hölscher D, Ludwig B, Möller RF, Fölster H (1997) Dynamic of soil chemical parameters in shifting agriculture in Eastern Amazon. Agric Ecosyst Environ 66:153–163. https://doi.org/10.1016/S0167-8809(97)00077-7
24. Juo ASR, Manu A (1996) Chemical dynamics in slash-and-burn agriculture. Agric Ecosyst Environ 58:49–60. https://doi.org/10.1016/0167-8809(95)00656-7
25. Palm CA, Swift MJ, Woomer PL (1996) Soil biological dynamics in slash-and-burn agriculture. Agric Ecosyst Environ 58:61–74. https://doi.org/10.1016/0167-8809(95)00653-2
26. Nath AJ, Sahoo UK, Giri K, Sileshi GW, Das AK (2020) Incentivizing hill farmers for promoting agroforestry as an alternative to shifting cultivation in Northeast India. In: Dagar JC, Gupta SR, Teketay D (eds) Agroforestry for degraded landscapes, Springer, Singapore, pp 425–444. https://doi.org/10.1007/978-981-15-4136-0_14
27. Roy PS, Kushwaha SPS, Murthy MSR, Roy A, Kushwaha D, Reddy CS, Behera MD, Mathur VB, Padalia H, Saran S, Singh S, Jha CS, Porwal MC (2012) Biodiversity characterization at landscape level: national assessment. Indian Institute of Remote Sensing, Dehradun, Uttarakhand, India
28. van Vliet N, Mertz O, Heinimann A, Langanke T, Pascual U, Schmook B, Adams C, Schmidt-Vogt D, Messerli P, Leisz S, Castella J-C, Jørgensen L, Birch-Thomsen T, Hett C, Bech-Bruun T, Ickowitz A, Vu KC, Yasuyuki K, Fox J, Padoch C, Dressler W, Ziegler AD (2012) Trends, drivers and impacts of changes in swidden cultivation in tropical forest-agriculture frontiers: a global assessment. Glob Environ Change 22:418–429. https://doi.org/10.1016/j.gloenvcha.2011.10.009
29. Tiwari BK, Pant RM (2018) Shifting cultivation: towards transformation approach: a draft report. National Institute of Rural Development and Panchayati Raj, North Eastern Regional Centre, Guwahati, Assam, India
30. Mukul SA, Herbohn J (2016) The impacts of shifting cultivation on secondary forests dynamics in tropics: a synthesis of the key findings and spatio temporal distribution of research. Environ Sci Policy 55:167–177. https://doi.org/10.1016/j.envsci.2015.10.005
31. Rai SC (2005) Apatani paddy-cum-fish cultivation: an indigenous hill farming system of North East India. Indian J Tradit Knowl 4:65–71
32. Chatterjee D, Deka BC, Patra MK, Thirugnanavel A, Borah TR, Kuotsu R, Ngachan SV (2012) Resilient shifting cultivation for sustainable soil-water-nutrient-plant continuum in hilly agriculture system of North Eastern India. In: Deka BC, Patra MK, Thirugnanavel A, Chatterjee D,

Borah TR, Ngachan SV (eds) Resilient shifting cultivation: challenges and opportunities. ICAR Research Complex for NEH Region, Nagaland Centre, Jharnapani, Medziphema, Nagaland, India, pp 104–109
33. Mylliemngap W (2021) Agrobiodiversity and natural resource management in traditional agricultural systems of Northeast India. Agrobiodivers Agroecol 1:1. https://doi.org/10.33002/aa010101
34. NITI Aayog (2018) Report of working group III: shifting cultivation: towards a transformational approach. New Delhi, India
35. Nath AJ, Brahma B, Lal R, Das AK (2016) Soil and jhum cultivation. In: Lal R (ed) Encyclopedia of soil science, 3rd edn. CRC Press, Boca Raton, FL, USA, pp 1273–1280. https://doi.org/10.1081/E-ESS3-120053748
36. Villa PM, Venâncio Martins S, de Oliveira Neto SN, Rodrigues AC, Pino HE, Kim D-G (2020) Shifting cultivation and agroforestry in the Amazon: premises for REDD+. For Policy Econ 118:102217. https://doi.org/10.1016/j.forpol.2020.102217
37. FAO (1995) Shifting cultivation in Bhutan: a gradual approach to modifying land use patterns: a case study from Pema Gatshel District, Bhutan. Community Forestry Case Study Series 11, Food and Agriculture Organization, Rome, Italy
38. Lal R (2015) Shifting cultivation versus sustainable intensification. Ref Modul Earth Syst Environ Sci https://doi.org/10.1016/B978-0-12-409548-9.09295-2
39. Barlow J, Lennox GD, Ferreira J, Berenguer E, Lees AC, Mac Nally R, Thomson JR, de Barros Ferraz SF, Louzada J, Oliveira VHF, Parry L, de Castro Solar RR, Vieira ICG, Aragão LEOC, Begotti RA, Braga RF, Cardoso TM, de Oliveira Jr RC, Souza CM Jr, Moura NG, Nunes SS, Siqueira JV, Pardini R, Silveira JM, Vaz-de-Mello FZ, Veiga RCS, Venturieri A, Gardner TA (2016) Anthropogenic disturbance in tropical forests can double biodiversity loss from deforestation. Nature 535:144–147. https://doi.org/10.1038/nature18326
40. FAO (2010) Global forest resources assessment 2010: main report. FAO Forestry Paper No. 163. Food and Agriculture Organization, Rome, Italy
41. Neyra-Cabatac NM, Pulhin JM, Cabanilla DB (2012) Indigenous agroforestry in a changing context: the case of the Erumanen ne Menuvu in southern Philippines. For Policy Econ 22:18–27. https://doi.org/10.1016/j.forpol.2012.01.007
42. Davidson EA, de Abreusa Sá TD, Reis Carvalho CJ, de Oliveia Figueriredo R, do Socorro A, Kato M, Kato OR, Ishida FY (2008) An integrated greenhouse gas assessment of an alternative to slash-and-burn agriculture in eastern Amazonia. Glob Change Biol 14:998–1007. https://doi.org/10.1111/j.1365-2486.2008.01542.x
43. Geist HJ, Lambin EF (2002) Proximate causes and underlying driving forces of tropical deforestation. Bioscience 52:143–150. https://doi.org/10.1641/0006-3568(2002)052[0143:PCAUDF]2.0.CO;2
44. Ramakrishnan PS (2016) Jhum: shifting agriculture. In: Lal R (ed) Encyclopedia of soil science, 3rd edn. CRC Press, Boca Raton, FL, USA, pp 1281–1284. https://doi.org/10.1081/E-ESS3-120016599
45. Biswal DK, Kumar S (2013) Shifting cultivation and policies of sustainable development: a meaningless obsession. J Econ Soc Dev 9:132–138
46. Mertz O (2009) Trends in shifting cultivation and the REDD mechanism. Curr Opin Environ Sustain 1:156–160. https://doi.org/10.1016/j.cosust.2009.10.002
47. Morton O, Borah JR, Edwards DP (2020) Economically viable forest restoration in shifting cultivation landscapes. Environ Res Lett 15:064017. https://doi.org/10.1088/1748-9326/ab7f0d
48. Nath AJ, Lal R, Das AK (2015) Grains for ecosystem carbon management in North East India. Curr Sci 109:1387–1389
49. Hossain MA (2011) An overview on shifting cultivation with reference to Bangladesh. Sci Res Essay 6:6509–6514. https://doi.org/10.5897/SRE11.1282
50. Verma PK, Kumar V, Chandra A, Thounaojam B (2017) Alternatives of shifting cultivation in North-eastern region of India. Rep Opin 9(12):1

Management of Soil Fertility and Crop Productivity Under Shifting Agriculture in Mizoram, North-East India: Ways Forward

Shri Kant Tripathi, Francis Q. Brearley, and Arun Jyoti Nath

Abstract Shifting agriculture, a traditional cultivation system, has been practiced in North-east India for centuries by the majority of the hill farming communities for their subsistence. Previously, this system of cultivation, with sufficiently prolonged fallow periods (i.e. about 20–30 years), was often remarkably productive and ecologically balanced. However, shortened fallow periods in recent times have become a common practice because of expanding populations, shrinking area for cultivation, and increased demand for food. Shorter fallow periods do not allow the soil to restore adequate fertility to support sufficient crop growth, leading to widespread concerns about ecological sustainability. This chapter aims to explore options to improve the efficiency of shifting cultivation systems of Mizoram state in North-east India by applying the principles of soil fertility management in natural ecosystems. This approach integrates ecological processes that support high productivity under mixed cropping by utilising diverse species niches and including different practical alternatives to improve shifting cultivation. For example, practicing inter-row cropping between contour hedgerows of nitrogen-fixing shrubs like *Flemingia macrophyla* and *Tephrosia candida*, promoting terrace cultivation on slopes, and use of anti-erosional plants for establishing agroforestry systems. Further application of innovative approaches such as amendment of litter from nearby forests, use of indigenous microbes as biofertilizers, and enhancement of fallow recovery rates would likely boost soil fertility and crop productivity in shifting cultivation. Based on our synthesis, we suggest promoting site-specific indigenous tree-based subsistence farming systems where shifting cultivation is the dominant land-use. These

S. K. Tripathi (✉)
Department of Forestry, Mizoram University, Tanhril, Aizawl, Mizoram 796009, India
e-mail: mzut041@mzu.edu.in

F. Q. Brearley
Department of Natural Sciences, Manchester Metropolitan University, Chester Street, Manchester M1 5GD, UK
e-mail: f.q.brearley@mmu.ac.uk

A. J. Nath
Department of Ecology and Environmental Sciences, Assam University, Silchar, Assam 788011, India

S. K. Tripathi and F. Q. Brearley (eds.), *Shifting Cultivation Systems*,
https://doi.org/10.1007/978-3-031-70388-1_16

systems can utilise the resources in an appropriate manner due to their varied growth and nutrient acquisition strategies, thus promoting agro-ecosystem development and improving productivity, diversity, and resilience.

Keywords Agroforestry · Agronomy · Soil fertility · Tree-based farming

1 Introduction

Forests in tropical regions are impacted by widespread deforestation primarily due to demand for agricultural land and economic expansion [1, 2]. Thus, evaluating land-use options for future sustainable development is a clear challenge. India launched the 'Green Revolution' in 1960s/70s to meet the growing demand for food by an increasing population from finite land resource, which was based on increased use of high yielding varieties, i.e. novel crop strains developed through selective breeding [3, 4], and increased the use of fertilisers to gain from these nutrient-responsive varieties. However, the application of commercial fertiliser in Indian tropical soils has led to decline in overall soil fertility due to a decrease in other plant nutrients and reduction in soil organic matter content under agriculture intensification, leading to eutrophication of rivers and lakes because of inefficient crop nutrient uptake [5, 6].

Shifting cultivation in North-east India (colloquially known as jhum) is a common practice of agriculture that sustains much of the rural population in this region (Fig. 1). Previously, the practice was sustainable because of prolonged fallow periods (ca. 20–30 years) that allowed the sites to recover fertility due to low population densities [7–9]. However, the current trend toward reduced fallow periods due to expanding population endangers its viability, leading to reduced crop yields and increased soil erosional losses that may be up to 60 Mg ha^{-1} yr^{-1} [10]. This has led to widespread concern from numerous stakeholders for the livelihood of the farming communities depending on it. Thus, government agencies in the region have implemented various policies with the aim of improving the livelihoods of the shifting cultivation farmers, many of whom remain poor [11]. Many policies implemented related to shifting cultivation were mostly focused on weaning away of the practice by replacing it with the other system of cultivation and were short-term (<5 years), and thus had limited success [11]. Therefore, sound policies based on high quality research outputs, particularly considering the sentiments of the people of the state, are required such that they can be widely accepted as cost-effective and eco-friendly harmonious options to supplement jhum cultivation rather than trying to replace the system of shifting cultivation by changing the tradition and culture of the society.

While the total area under shifting cultivation across all states of North-east India has reduced to some extent over recent years, it has continued to grow in states like Mizoram and Manipur. In 2003, the Task Force on Shifting Cultivation set up by the Government of India, reported [cited in [12]], an estimated area of 1.73×10^6 ha under this practice in North-east India during the period 1987–97 with an estimated 620,000 families dependent on shifting cultivation in the same time period. According

Fig. 1 Recently planted shifting cultivation field in Mizoram, North-east India

to a more recent NITI Aayog report [12], about 8.5×10^5 ha of land in the Indian eastern Himalayan region is still used for shifting cultivation. Most of the studies available on shifting cultivation systems in tropical belts have been carried out on lands having relatively flat or gentle slopes. Further, shifting cultivation in well-studied Meghalaya state has been focused on the areas having gentle to moderate slopes [7]. Therefore, the conclusions and recommendations for the management of shifting cultivation systems or any other alternative forms of agriculture for those regions needs to be carefully evaluated for their application in the specific context of Mizoram where the majority of the landscapes are characterized by steep slopes [13]. Of the total planimetric land area of the state (c. 21,087 km^2), only about one quarter has gentle slopes (10°–33°), whilst the remaining three-quarters have slopes steeper than 33° where the practice of shifting cultivation is being carried out by about half of all households. Hence, the solutions to sustain shifting cultivation in a state like Mizoram will be site-specific and distinctive from other areas of North-east India and elsewhere. This chapter, therefore, suggests viable options to sustain soil fertility and crop productivity in the steeply sloping regions of North-east India with a particular focus on the state of Mizoram.

2 Ecological Studies on Nutrient Cycling Under Shifting Cultivation

Ramakrishnan [7] has summarised a range of ecological studies on shifting cultivation carried out in North-east India from different agricultural land-uses and climatic locations. Further, Toky and Ramakrishnan [14], have conducted ecological studies at the low elevation Burnihat site in Meghalaya with moderate slopes (20°–40°) that resembles typical climatic conditions in Mizoram. Their study compared the soil biogeochemistry and economic yields of three fallow ages (i.e. 5-, 10- and 30-years) on replicated sites with comparable slopes. The authors reported that the total economic yields on an annual basis were more than two-fold more in the long fallow sites (i.e. 10- and 30-years) compared to that of the short fallow site (i.e. 5-year-old) [14]. In contrast, in another study, the run-off losses (per year) of soluble nutrients from surface soil were considerably (ca. 40%) greater in the 5-year-old fallows than in the long fallows (10- and 30-years) [15]. This result suggests that the loss of soil nutrients from the short fallow sites (i.e. 5 years) could possibly be one of the major causes for the decrease in crop productivity in shifting cultivation sites [10]. Further studies in these sites showed considerable reduction (ca. 25%) in soil organic carbon (C) and nitrogen (N) concentrations (in the top 7 cm depth) in 5-year-old fallow compared to prolonged fallows (10- and 30-years). However, the pool sizes of C and N were unaffected, which may be because of increased bulk density in the 5-year fallow site due to greater reduction in organic matter content compared to the long fallow sites [16]. Similarly, concentrations of available phosphorus (P) and potassium (K) were also considerably (i.e. two to three times) lower in the 5-year-old fallow compared to the prolonged fallow ages. Furthermore, the same authors found that burning of vegetation decreased C concentrations in surface soil in the 10- and 30-year old fallows by around 15%, which was unaffected in the short fallow (5-year-old) site. This was attributed to less intense fires due to decreased above-ground biomass and litter mass in the shortest fallow length site [16]. Tawnenga et al. [17] reported the extended cropping period for the second year following burning in Mizoram by the effective management of cropping through suitable crop choices. The authors further reported that total ecosystem productivity (including production of crops and weeds) in 6-year-old fallow decreased by ca. 45% and 22% in two consecutive years after burning as compared to 20-year-old fallow. In a follow-up paper, Tawnenga et al. [18] also reported significant depletion in the amount of soil organic C, total N, extractable P, and exchangeable cation pools in the short fallow compared to the longer one. Further, the authors noticed a gradual decrease in soluble nutrients at both the sites during each successive cropping phase.

Later, Grogan et al. [13] summarized these studies with the following conclusions for sustainable crop production practices under shifting cultivation in Mizoram. They concluded that: (a) a minimum of 10-years fallow length would be required for the sufficient recovery of the site to support economically viable and ecologically balanced shifting cultivation in North-east India; (b) shortened fallow period of 5-years or less would considerably reduce soil fertility (including the availability of P and other base cations) because of substantially enhanced sediment run-off losses rather than soluble nutrient losses; (c) reduced supply of plant-available nutrients in shortened fallow lands (ca. 5-years) are because of depleted soil C pools (i.e., organic matter content) in younger fallows; and (d) extending the site use period to second- and third-year cropping would be the most promising option to reduce the area required for burning each year under shifting cultivation. However, the management of soil fertility and weeds under shifting cultivation in the second year onwards are more labor intensive, and hence, the farmers are not continuing this practice past the second year. Numerous review articles available on this topic have clearly indicated that the impacts of shortening fallow periods needs further strengthening based on our deeper understanding of shifting cultivation systems [7, 9, 13, 19].

3 Management of Soil Fertility Under Shifting Cultivation on Steep Slopes of Mizoram

As noted earlier, Mizoram is characterised by steep slopes where soil erosional losses under shifting cultivation are larger than other places, and thus require distinctive solutions considering the steep slopes of the state [13]. Considering the ecological and environmental understanding of the problem, Grogan et al. [13] suggested several options for improving the efficiency of shifting agriculture in Mizoram, most of which still remain relevant in today's context (Table 1). Among them, the most promising choices for improving the efficiency of shifting cultivation in this region are, for example, supplementation of water and nutrients, optimization of crop choices, incorporation of socially and ecologically important native tree species, extension of the period for site use, reduction in minimum required fallow period, and increased control on biomass burns and their environmental impacts (Table 1). Further studies on weed biology and weed-crop interactions would also assist in enhancing crop productivity of shifting cultivation practices. Further, the introduction of N-fixing plants (i.e., *Flemingia macrophylla* and *Tephrosia candida*; both Fabaceae; Fig. 2) along with the selection of other suitable plant species that can grow well in nutrient-deficient shifting cultivation sites would be beneficial in overcoming the problem of depletion of critical nutrients like N and P. In addition to leaf litter, *Tephrosia candida* root necromass can add nutrients to soil [20]. Additionally, the application of manure and crop and weed residues such as organic mulch can be trialed [21]. Some farmers are using either one or the other trails for improving the soil fertility and crop productivity under shifting cultivation in Mizoram. However, a detailed innovative scientific model for different ecological regions based high quality science would benefit the farmers at large for sustainable crop production practices in the region.

In addition to the recommendations of Grogan et al. [13] as itemised in Table 1, there is much to be learned from the science of ecological restoration and its synergies with agro-ecology. For example, studies in the Neotropics have shown how agricultural waste, such as orange peelings or coffee waste, can be applied at high rates and increase the speed at which secondary forests (fallows) accumulate biomass and soil nutrients [22]. In Mizoram, Wapongnungsang [23] added leaf litter (5 Mg ha^{-1}) to jhum fields and showed how this may increase rice (*Oryza sativa*; Poaceae) yield. Basic fertilisation experiments are surprisingly rare in shifting cultivation fields. Whilst we might expect P to be an essential limiting nutrient, the role of N should not be underestimated, nor should the influence of soil acidity [24]. Indeed, further studies on the full field-scale or catchment scale nutrient budgets would be beneficial for developing appropriate nutrient management strategies [25, 26].

Table 1 Important research options to consider in order to improve shifting cultivation practices in Mizoram

Potential options	Research required
Nutrient and water supplementation	(i) To better understand the relative importance of water and nutrients (particularly N, P, K) in influencing crop growth and economic yields in this region (ii) To find out key site factors (like slope, aspect, soil type, and climate) that affect crop growth and production in the region (iii) To better understand the optimum amount of organic matter inputs needed to increase crop growth. Further, the source of these inputs (e.g. composted weeds and crop residues or green manure) from the site or from nearby areas (iv) To understand the potential of applying commercial fertilisers to increase crop yields and determine application methods to minimise nutrient loss (v) ** To determine the potential of food wastes/mulches/forest leaf litter/rock phosphate in enhancing crop yields (vi) ** To ascertain full nutrient budgets under contrasting conditions (i.e. differing fallow ages, slope gradients, climate, etc.) (vii) ** To assess nutrient limitation in soil and nutrient requirements of different crops
Crop choice and extending the site use period	(i) To better understand the optimum crop species that can be grown together and to determine the maximum length of fallow period for subsistence and improved economic yields of crops under shifting cultivation in the region (ii) To work out the role of local site factors (e.g. slope, aspect, soil type, and climate) on crop productivity and find out the nearby markets for choice of crop produce (iii) To develop wild plants as appropriate crops for the region, and to find suitable perennial crops appropriate to sustain economic yields over successive years (iv) To investigate weed biology for developing appropriate techniques to suppress weeds in the 2nd and 3rd year of cropping (v) To determine ecologically effective and economically feasible slope management options to reduce soil losses (vi) ** To understand the feasibility of combining various leguminous crops (annual/perennial) along with cereals and vegetables

(continued)

Table 1 (continued)

Potential options	Research required
Enhancing fallow recovery rates	(i) To better understand the role of N-fixing and other native species for their possibility to be planted on sites to enhance vegetation succession following cropping (ii) To work out suitable cover species that can be planted after cropping to protect the soils from their erosional losses (iii) To determine the effect of addition of fertilizer on enhancing vegetation regeneration, and its economic feasibility, and to understand nutrient limitations, if any (iv) ** To know which tree species could be planted to enhance fallow recovery (and also provide economic benefits) (v) ** To understand the role of early regenerating fast-growing species like broom grass (*Thysanolaena latifolia*; Poaceae), cassava (*Manihot esculenta*; Euphorbiaceae) and neem (*Azadirachta indica*; Meliaceae) in accelerating fallow recovery rate by rapidly adding considerable amounts of organic matter
Improvement of fire management and its environmental impacts	(i) To know the potential for improving fire management, for example, (a) influence the proportion of land burned each year but not cultivated and (b) control the extent of burns using fire breaks effectively (ii) To know the potential of bamboo varieties and other fast-growing native trees to be effectively utilised as windbreaks to minimise wind- and water-borne soil, ash, and soluble nutrient losses (iii) To know the effect of mulching with organic matter residues after the burn on crop cultivation (iv) ** To understand the effect of inoculation of topsoil of adjoining forests after burning (v) ** To determine the relative emissions of CO_2 from the burning of biomass and its decomposition

(continued)

Table 1 (continued)

Potential options	Research required
Nitrogen-fixing shrub hedgerows	(i) Assess temporal patterns of availability of potential growth limiting nutrients and whole farm nutrient budgets over at least a decade (ii) Determine the period of crop productivity maintenance in continuous cultivation sites by assessing individual and overall crop productivity and yield patterns (iii) Assess the economic viability of this practice with financial and labour investment analyses (iv) Test the suitability of native and exotic nitrogen-fixing shrub species on different slopes, soil types and climatic conditions (v) Determine crop and non-crop interactions and their influence on growth performances
Slope terracing	(i) Assess the economic and ecological feasibility of terrace slope angle, height, and terrace walls using contour fences of bamboo or tree stem poles (ii) Determine the suitability of various crops and wild plant species for optimising terraced production
Soil microbiology**	(i) ** To understand how aspects of soil microbiology (composition and function) recover during vegetation regeneration (ii) ** To understand plant–microbe interactions particularly early regenerating plant species (iii) ** To further prospect for micro-organisms with plant growth promoting properties and conduct growth trials on various crops for their effectiveness
Cultural conditions	(i) ** To increase understanding of cultural conditions that tie farmers to shifting cultivation and the factors that may lead to acceptance of alternative agricultural options, such as agroforestry systems (ii) ** To assess the initial investments and cost–benefit analysis in establishing farming practices other than shifting cultivation

Modified from Grogan et al. [13] to include several additional options (marked with **)

The role of shifting cultivation in influencing soil micro-organisms and their subsequent succession during fallows is far from clear; this is an obvious research gap as we know remarkably little about the microbial communities found in shifting cultivation fields and fallows. Microbial inoculants are gaining in interest and importance globally [27]. Pot experiments based on microbes isolated from shifting cultivation sites indicated an increase in rice growth by up to around 25% when inoculated with plant-growth promoting (PGP) micro-organisms [28] and with a similar magnitude

Fig. 2 *Tephrosia candida* hedgerow in Muallungthu village, Mizoram, North-east India

of increase in field experiments [28–30]. Further prospecting for PGP producing micro-organisms would be of benefit to increase the number of possible options [31].

Various options to augment shifting cultivation with continuous cropping have been suggested, for example, cropping within two hedgerows of N-fixing shrubs raised alongside of the slopes to reduce soil erosional losses (Fig. 2), terracing of slopes for minimising the slope effect, raising of anti-erosional plants having profuse rooting, and bamboo forest farms etc. [10]. The role of bamboo in stabilising nutrient cycling in the fallow lands of shifting cultivation has been studied widely. For example, *Melocanna baccifera* (Poaceae) has been reported to potentially increase microbial activities in fallow lands, which consequently enhance soil C and N in the fallows [32]. In the dry tropical region in India, bamboos have been reported to develop extensive adventitious roots systems that outcompete herbs for exploiting soil nutrients and water from oligotrophic soils under water deficient conditions [33, 34]. Planting of valuable trees such as tree bean (*Parkia timoriana*; Fabaceae) or agarwood (*Aquilaria malaccensis*; Thymelaeaceae) in the fallows have been considered [35, 36], but the role of these in influencing soil properties is little known.

However, some *Macaranga* (Eurphorbiaceae) trees appear to have fallow nutrient-enriching properties, as farmers in Nagaland retain these trees amongst the rice crop thus allowing them to dominate the fallow canopy after cessation of cultivation [37].

4 The Ways Forward

Long fallow cycles were possible earlier because local human population pressure on the environment was not as intense, and thus forested land availability did not limit the shifting cultivation cycle [38]. With the shortening of the jhum cycle, there has been an impact on the energy ratio of outputs and inputs, which ultimately affects the farmers' economic returns (Table 2) [39].

Over the years, shifting cultivation lands have also been targeted by multiple government agencies for conversion to cash crops like tea (*Camellia sinensis*; Theaceae), coffee (*Coffea arabica*; Rubiaceae), rubber (*Hevea brasiliensis*; Euphorbiaceae) and timber species [40]. The government of Mizoram has recently started promoting plantations of species of palm (*Areca catechu*; Arecaceae) and oil palm (*Elaeis guineensis*; Arecaceae). This reallocation of land for cultivation of cash-crops is an additional major cause contributing to the reduction in total land available for shifting cultivation and the consequent shortening of the fallow periods [8]. However, the existing government policies do not provide alternative income opportunities for farming communities. Further, farmers do not have greater tenurial rights that also enhances pressure on shifting cultivation lands. As a result, shifting cultivation is becoming less sustainable in this region. Moreover, replacing shifting cultivation systems with settled agriculture in the region has also been proved to be unsustainable because of its geography. Additionally, introduction of modern agricultural intensification has also negatively affected the region, such as declines in soil fertility and drying of water sources [12].

On the other hand, some of the Indigenous communities in this region have successfully transformed shifting cultivation to traditional agroforestry systems [41]. With the domestication of woody perennials such as *Albizia procera* (Fabaceae), *Gmelina arborea* (Lamiaceae), *Mangifera indica* (Anacardiaceae), *Neolamarckia cadamba* (Rubiaceae), *Parkia timoriana* and *Phyllanthus emblica* (Phyllanthaceae) as an integral component, these systems have been transformed to fall under the

Table 2 Energy balance in jhum fallows in North-east India

Fallow period (years)	Energy (Mj ha^{-1} yr^{-1})		
	Input	Output	Output/Input ratio
30	1665	56,760	34.1
10	1181	56,600	47.9
5	510	23,860	46.7

Source Ramakrishnan [39]

category of landscape-scale mitigation schemes by reducing emissions from deforestation and degradation—REDD+. Additionally, traditional land-use systems practiced in the region further deliver a number of benefits and services to the society. For instance, providing fuel wood, timber, medicines, replenishing soil nutrients, binding soil to reduce erosional losses, conservation of soil moisture, and improving overall crop yield. Reports from southern Assam [42, 43] have revealed that this transformation of shifting cultivation developed over the last century as the availability of land decreased, and there is a need for increased household income. This thus appears to be an evolution as part of the culture of Indigenous peoples of the region, noting that agricultural practices are closely linked with their sociocultural practices and beliefs. Thus, it is essential to remain cognisant instead of forcing communities to wean away from traditional systems of cultivation and natural resource management through region-specific policy formulation and provision of fiscal support to promote community-specific best management practices.

In line with the action points proposed by NITI Aayog [12], we suggest any future programmes must aim to: (a) undertake careful diagnosis of the shifting cultivation situation, (b) document and consolidate the lessons on the magnitude of the concerns raised; (c) identify viable best practices with potential for upscaling, and (d) ascertain to what extent and which 'co-benefits' could be delivered to the hill farmers. In conclusion, alternatives to shifting cultivation practices in Mizoram have been tested, but there has been no successful experience in widespread farmer adoption or development of innovative concepts like nature-based farming and forest landscape restoration. Therefore, identifying and promoting more sustainable innovations for low-input and low-technology land-use practices that are socially and culturally acceptable may ensure better livelihoods, and contribute to environmental conservation. Therefore, these choices are required to be properly evaluated and examined before they can be recommended as part of land-use planning initiatives for their adoptions. Additionally, while recommending improvement alternatives to shifting cultivation practice, precautionary measures should be taken that the research focuses on ensuring realistic, feasible, economically possible, and culturally appropriate measures. Therefore, the success of the practical solutions developed will depend on incorporation and integration of economic, ecological, and social components of shifting cultivation in the region [7].

Further, care should be taken that the research outputs and technologies developed for their application in shifting cultivation system must be acceptable and accessible to poor farmers. In addition, an active extension education programme addressing landholders' need and social constraints faced by them would be required through field demonstration plots established across the region for successful society-oriented results [44]. Due to the current concerns regarding shifting cultivation, it is crucial and urgent to test, formalize, and promote sustainable innovations that will positively impact both livelihoods and environmental conservation in Mizoram.

References

1. Lewis SL, Edwards DP, Galbraith D (2015) Increasing human dominance of tropical forests. Science 349:827–832. https://doi.org/10.1126/science.aaa9932
2. Edwards DP, Socolar JB, Mills SC, Burivalova Z, Koh LP, Wilcove DS (2019) Conservation of tropical forests in the anthropocene. Curr Biol 29:R1008–R1020. https://doi.org/10.1016/j.cub.2019.08.026
3. Evenson RE, Gollin D (2003) Assessing the impact of the Green Revolution, 1960 to 2000. Science 300:758–762. https://doi.org/10.1126/science.107871
4. Eliazer Nelson ARL, Ravichandran K, Antony U (2019) The impact of the Green Revolution on indigenous crops of India. J Ethn Food 6:8. https://doi.org/10.1186/s42779-019-0011-9
5. Singh KP, Tripathi SK (2000) Impact of environmental nutrient loading on the structure and functioning of terrestrial ecosystems. Curr Sci 79:316–323
6. Stevens CJ (2019) Nitrogen in the environment. Science 363:578–580. https://doi.org/10.1126/science.aav8215
7. Ramakrishnan PS (1992) Shifting agriculture and sustainable development: an interdisciplinary study from North-eastern India, UNESCO Man and the Biosphere Series, vol 10. UNESCO and Parthenon Publishing Group, Paris, France
8. Kerkhoff E, Sharma E (2006) Debating shifting cultivation in the Eastern Himalayas: farmers' innovations as lessons for policy. International Center for Integrated Mountain Development, Kathmandu, Nepal
9. Bruun TB, de Neergaard A, Lawrence D, Ziegler A (2009) Environmental consequences of the demise in swidden cultivation in Southeast Asia: carbon storage and soil quality. Hum Ecol 37:375–388. https://doi.org/10.1007/s10745-009-9257-y
10. Choudhury B, Nengzouzam G, Ansari MA, Islam A (2022) Causes and consequences of soil erosion in northeastern Himalaya, India. Curr Sci 122:772–789. https://doi.org/10.18520/cs/v122/i7/772-789
11. Tripathi SK, Vanlalfakawma DC, Lalnunmawia F (2017) Shifting cultivation on steep slopes of Mizoram, India: impact of policy reforms. In: Cairns M (ed) Shifting cultivation policies: balancing environmental and social sustainability. CABI, Wallingford, UK, pp 393–413. https://doi.org/10.1079/9781786391797.0393
12. NITI Aayog (2018) Report of working group III: shifting cultivation: towards a transformational approach. New Delhi, India
13. Grogan P, Lalnunmawia F, Tripathi SK (2012) Shifting cultivation in steeply sloped regions: a review of management options and research priorities for Mizoram state, Northeast India. Agrofor Syst 84:163–177. https://doi.org/10.1007/s10457-011-9469-1
14. Toky OP, Ramakrishnan PS (1981) Cropping and yields in agricultural systems of the north-eastern hill region of India. Agro-Ecosyst 7:11–25. https://doi.org/10.1016/0304-3746(81)90012-3
15. Toky OP, Ramakrishnan PS (1981) Run-off and infiltration losses related to shifting agriculture (jhum) in northeastern India. Environ Conserv 8:313–321. https://doi.org/10.1017/S0376892900028125
16. Ramakrishnan PS, Toky OP (1981) Soil nutrient status of hill agro-ecosystems and recovery pattern after slash and burn agriculture (Jhum) in north-eastern India. Plant Soil 60:41–64. https://doi.org/10.1007/BF02377111
17. Tawnenga SU, Tripathi RS (1996) Evaluating second year cropping on jhum fallows in Mizoram, north-eastern India—phytomass dynamics and primary productivity. J Biosci 21:563–575. https://doi.org/10.1007/BF02703219
18. Tawnenga SU, Tripathi RS (1997) Evaluating second year cropping on jhum fallows in Mizoram, north-eastern India: soil fertility. J Biosci 22:615–625. https://doi.org/10.1007/BF02703399
19. Craswell ET, Sajjapongse A, Howlett DJB, Dowling AJ (1997) Agroforestry in the management of sloping lands in Asia and the Pacific. Agrofor Syst 38:121–137. https://doi.org/10.1023/A:1005960612386

20. Wapongnungsang, Hauchhum R, Tripathi SK (2018) Dry matter, carbon and nitrogen inputs from Tephrosia candida hedgerows in different fallow lands following shifting cultivation in Mizoram. Multilog Sci 7:267–271
21. Lalremsang P, Upadhyaya K, Sahoo UK, Singson L (2022) Effect of legume leaf mulch and fertilizer on soil quality and rice yield for small scale production. Acta Ecol Sin 43:861–868. https://doi.org/10.1016/j.chnaes.2022.12.006
22. Choi JJ, Treuer TLH, Werden LK, Wilcove DS (2018) Organic wastes and tropical forest restoration. Trop Conserv Sci 11:1–5. https://doi.org/10.1177/1940082918783156
23. Wapongnungsang (2018) Impact of fallow periods and treatments on the pattern of recovery of soil fertility and plant productivity in shifting cultivation sites in Aizawl district, Mizoram. Ph.D. Thesis, Mizoram University, Aizawl, Mizoram, India
24. Lungmuana RY, Singh SB, Saha S, Soni JK, Shakuntala I (2023) Response of lime and phosphorus application on groundnut growth, yield and soil enzyme activities in acidic soil of North Eastern India. Commun Soil Sci Plant Anal 54:1616–1626. https://doi.org/10.1080/00103624.2023.2188071
25. Hölscher D, Möller RF, Denich M, Fölster H (1996) Nutrient input-output budget of shifting agriculture in Eastern Amazonia. Nutr Cycl Agroecosyst 47:49–57. https://doi.org/10.1007/BF01985718
26. Borggaard OK, Gafur A, Petersen L (2003) Sustainability appraisal of shifting cultivation in the Chittagong Hill Tracts of Bangladesh. Ambio 32:118–123. https://doi.org/10.1579/0044-7447-32.2.118
27. Trivedi P, Mattupalli C, Eversole K, Leach JE (2021) Enabling sustainable agriculture through understanding and enhancement of microbiomes. New Phytol 230:2129–2147. https://doi.org/10.1111/nph.17319
28. Banerjee A, Bareh DA, Joshi S (2017) Native microorganisms as potent bioinoculants for plant growth promotion in shifting agriculture (*jhum*) systems. J Soil Sci Plant Nutr 17:127–140. https://doi.org/10.4067/S0718-95162017005000010
29. Wapongnungsang, Ovung, EY, Upadhyay KK, Tripathi SK (2021) Soil fertility and rice productivity in shifting cultivation: impact of fallow lengths and soil amendments in Lengpui, Mizoram northeast India. Heliyon, 7:e06834. https://doi.org/10.1016/j.heliyon.2021.e06834
30. Giri K, Mishra G, Suyal DC, Kumar N, Doley B, Das N, Baruah RC, Bhattacharyya R, Bora N (2023) Performance evaluation of native plant growth-promoting rhizobacteria for paddy yield enhancement in the jhum fields of Mokokchung, Nagaland, North East India. Heliyon 9:e14588. https://doi.org/10.1016/j.heliyon.2023.e14588
31. Hnamte R, Lalruatsanga H, Lalfakzuala R (2021) Isolation, identification and characterization of phosphate solubilizing bacteria from jhum field of Mizoram, India. Geomicrobiol J 38:924–933. https://doi.org/10.1080/01490451.2021.1980922
32. Hauchhum R, Tripathi SK (2017) Rhizosphere effects of *Melocanna baccifera* on soil microbial properties under different fallow phases following shifting cultivation. Int J Plant Soil Sci 17:34493. https://doi.org/10.9734/IJPSS/2017/34493
33. Tripathi SK, Singh KP (1994) Productivity and nutrient cycling in recently harvested and mature bamboo savannas in the dry tropics. J Appl Ecol 31:109–124. https://doi.org/10.2307/2404604
34. Tripathi SK, Singh KP, Singh PK (1999) Temporal changes in spatial pattern of fine-root mass and nutrient concentrations in Indian bamboo savanna. Appl Veg Sci 2:229–238. https://doi.org/10.2307/1478986
35. Ovung EY, Loya B, Brearley FQ, Tripathi SK (2021) Ethnic uses of *Parkia timoriana* (Fabaceae) and their significance to the Lotha tribes of Nagaland, North-east India. Vegetos 34:77–85. https://doi.org/10.1007/s42535-020-00178-7
36. Giri K, Jayaraj RSC, Mishra G, Bordoloi (2020) *Aquilaria malaccensis* Lamk.; a potential bioresource for shifting cultivation management, livelihood generation and climate change mitigation in North East India. Proc Indian Natl Sci Acad 86:1287–1289. https://doi.org/10.16943/ptinsa/2019/49705

37. Changkija S, Thakuria D, Cynthia A (2023) Traditional use of *Macaranga* trees for soil fertility by Naga shifting cultivators in northeast India. In: Cairns M (ed) Farmer innovations and best practices by shifting cultivators in Asia-Pacific. CABI, Wallingford, UK, pp 557–570. https://doi.org/10.1079/9781800620117.0026
38. Nath AJ, Reang D, Sileshi GW (2022) The shifting cultivation juggernaut: an attribution problem. Glob Chall 6:2200051. https://doi.org/10.1002/gch2.202200051
39. Ramakrishnan PS (1991) Ecology and sustainable development. National Book Trust, New Delhi, India
40. DTE Staff (2018) Govt needs to address policy ambiguities affecting shifting cultivation: NITI Aayog. https://www.downtoearth.org.in/news/agriculture/govt-needs-to-addresspolicy-ambiguities-affecting-shifting-cultivation-niti-aayog-61519. Accessed 2 Nov 2021
41. Nath AJ, Sahoo UK, Giri K, Sileshi GW, Das AK (2020). Incentivizing hill farmers for promoting agroforestry as an alternative to shifting cultivation in Northeast India. In: Dagar JC, Gupta SR, Teketay D (eds) Agroforestry for degraded landscapes. Springer, Singapore, pp 425–444. https://doi.org/10.1007/978-981-15-4136-0_14
42. Reang D, Hazarika A, Sileshi GW, Pandey R, Das AK, Nath AJ (2021) Assessing tree diversity and carbon storage during land use transitioning from shifting cultivation to indigenous agroforestry systems: implications for REDD+ initiatives. J Environ Manage 298:113470. https://doi.org/10.1016/j.jenvman.2021.113470
43. Reang D, Nath AJ, Sileshi GW, Hazarika A, Das AK (2022) Post-fire restoration of land under shifting cultivation: a case study of pineapple agroforestry in the sub-Himalayan region. J Environ Manage 305:114372. https://doi.org/10.1016/j.jenvman.2021.114372
44. Cairns M, Garrity DP (1999) Improving shifting cultivation in Southeast Asia by building on indigenous fallow management strategies. Agrofor Syst 47:37–48. https://doi.org/10.1023/A:1006248104991

The manufacturer's authorised representative in the EU is Springer Nature Customer Service Centre GmbH, Europaplatz 3, 69115 Heidelberg, Germany. If you have any concerns regarding our products, please contact ProductSafety@springernature.com

Printed and bound by CPI Group (UK) Ltd, Croydon, CR0 4YY
01/10/2026
02225910-0003